RESIDENTIAL OIL BURNERS

Second Edition

RESIDENTIAL OIL BURNERS

Second Edition

HERB WEINBERGER

DELMAR

™

THOMSON LEARNING

Australia   Canada   Mexico   Singapore   Spain   United Kingdom   United States

DELMAR

THOMSON LEARNING

Residential Oil Burners, Second Edition
by
Herb Weinberger

Cover Photographs: (top) LEDV Boiler with Beckett AFII Burner, courtesy of Burnham Corporation; (middle) EZPro Oil Burner, courtesy of Carlin Combustion Technology; (bottom) Bacharach Oil Burner Combustion Efficiency Kit, courtesy Bacharach, Inc.

Business Unit Director: Alar Elken	**Developmental Editor:** Catherine Wein	**Executive Marketing Manager:** Maura Theriault
Executive Editor: Sandy Clark	**Executive Production Manager:** Mary Ellen Black	**Channel Manager:** Mona Caron
Acquisitions Editor: Vernon Anthony	**Production Editor:** Ruth Fisher	**Marketing Manager:** Kasey Young

COPYRIGHT © 2001 by Delmar, a division of Thomson Learning, Inc. Thomson Learning™ is a trademark used herein under license.

Printed in Canada
3 4 5 XXX 05 04 03

For more information contact Delmar,
3 Columbia Circle, PO Box 15015,
Albany, NY 12212-5015.

Or find us on the World Wide Web at
http://www.delmar.com

For permission to use material from this text or product, contact us by
Tel   (800) 730-2214
Fax   (800) 730-2215
www.thomsonrights.com

Library of Congress Cataloging-in-Publication Data

Weinberger, Herb.
   Residential oil burners / Herb Weinberger.—2nd ed.
      p.   cm.
   Includes index.
      ISBN 0-7668-1828-4 (text : alk. paper)—
   ISBN 0-7668-1829-2 (IG : alk. paper)
      1. Oil burners—Maintenance and repair.   I. Title.
TH7466.06 W45 2001
697′.044—dc21   00-024024

NOTICE TO THE READER

Publisher does not warrant or guarantee any of the products described herein or perform any independent analysis in connection with any of the product information contained herein. Publisher does not assume, and expressly disclaims, any obligation to obtain and include information other than that provided to it by the manufacturer.

The reader is expressly warned to consider and adopt all safety precautions that might be indicated by the activities herein and to avoid all potential hazards. By following the instructions contained herein, the reader willingly assumes all risks in connection with such instructions.

The Publisher makes no representation or warranties of any kind, including but not limited to, the warranties of fitness for particular purpose or merchantability, nor are any such representations implied with respect to the material set forth herein, and the publisher takes no responsibility with respect to such material. The publisher shall not be liable for any special, consequential, or exemplary damages resulting, in whole or part, from the readers' use of, or reliance upon, this material.

To my wife, Leah

BRIEF CONTENTS

CONTENTS

PREFACE

This is the second edition of *Residential Oil Burners*. The first edition was published in 1992, but the basic purpose for writing the book is still the same. It is intended for use in conjunction with either a trade apprenticeship program or a vocational program at the secondary or post secondary level. The material in the text is unique in that no other single book on this subject contains the information found here. The conversational style of writing will make this an easy read for the students yet it still covers all of the technical aspects of the material essential to the training of service technicians.

New to this edition are the advances in oil burner technology from the United States and Europe, fully automatic space heaters from Japan, electronic ignition systems, new primary controls, and updated wiring and piping diagrams. All this and the basics, which are essentially unchanged, as well. Nothing from the first edition has been deleted. This increases the value of this text because it will become an ideal reference for future service personnel who occasionally run into a 20th-century heating unit problem.

The author wishes to thank the equipment manufacturers who contributed materials to this work. Their dedication to the industry is manifested in the continuing improvement in oil-fired heating equipment. Thanks to all the teachers, throughout the United States and Canada who have selected *Residential Oil Burners* as the text for their programs. Their comments and suggestions have been incorporated in this second edition. Thanks to all the professional educators who reviewed portions of this work, especially my ex-student, who is presently teaching at A. E. Smith Vocational High School, Mr. Pedro (Pete) Gonzalez.

This text is based on twenty-five years of personal experience working as an oil burner service and installation mechanic in and around New York City. An additional seventeen years as a teacher in a variety of programs, listed below, are also included. My fondest wish is that this text becomes part of every service technician's equipment.

- Alfred E. Smith Vocational High School
- New York City Technical College—Apartment House Institute
- New York City Technical College—Continuing Education
- Housing Conservation Coordinators of New York City
- New York City Transit Authority Heating Program
- New York City Board of Education Fireman Training Program
- New York City Housing Authority High School Training Program
- New York City Department of Air Resources Air Pollution Training Program

Herb Weinberger

CONTRIBUTORS

The following companies contributed materials to this text:

Adams Manufacturing Company
Aero Environmental Limited
Amtrol, Inc.
Bacharach, Inc.
R. W. Beckett Corp.
Bell and Gossett
Bock Water Heaters Inc.
Burnham Corporation
Cardinal Fibreglass Industries
Carlin Combustion Technology, Inc.
Danfoss Manufacturing Co. Ltd.
Delavan Inc.
Dornback Furnace and Foundry Company
The Field Controls
Honeywell
Lynn Products Company
McDonnell Miller
Monarch Manufacturing Works, Inc.
Riello Corporation of America
Scully Signal Company
Suntec Industries, Inc.
Tjernlund Products, Inc.
Toyotomi U.S.A. Inc.
Ultimate Engineering Corp.
Watts Regulator Company
Wayne Home Equipment
Webster Heating Products
Weil McLain
Z-FLEX

INTRODUCTION

Millions of homes depend on oil-fired heating equipment to maintain comfort and warmth regardless of how cold the weather conditions are outside. If you have ever had the experience of waking up to a cold house on a winter morning you can appreciate the need for trained oil burner service technicians. This book is designed to assist in the training of people interested in working in this trade. It should be used in conjunction with hands-on experience, either in a vocational–technical center shop or an industry apprenticeship program.

The materials in the text are arranged to supply the information necessary to perform the duties of an oil burner service technician. Each chapter deals with a specific portion that explains in detail the function of the system or parts involved. Included are the oil burners, the heating systems, basic electricity, controls and control circuit wiring, troubleshooting methods, production of domestic hot water, annual tune-up procedures, and efficiency testing. Perhaps a brief explanation and history of the industry will help get you started.

Service technicians are employed either by fuel oil dealers, local service contractors, or they are self-employed "freelance" contractors handling service calls for several dealers. Very often they are the only representatives of the company that the customer ever sees. Service personnel must have certain qualities that will ensure their success in repairing the equipment and satisfying the customer. The qualities needed are to:

1. Be friendly and courteous. There are times when you will come face-to-face with irate customers. It is up to you to calm them down and get the equipment repaired. Remember, you are in their home. Treat them as you would want someone to treat you in your home.
2. Make a clean, neat appearance. The company you work for may supply a uniform. It is up to you to keep it clean. Work carefully; clean up any dirt you may create and dispose of it outside of the building.
3. Develop your mechanical skills. Oil burners are not very complicated but to take them apart or to replace parts requires manual dexterity. You will quickly find out if you can do this type of work in the hands-on portion of your training.

4. Be dependable. Get to your job or call on time. Servicing oil burners will require that you work in some bad weather—cold, snow, heavy rain. You cannot stay home if it is snowing.

5. Take care of your equipment. Most companies will give you a service van to travel in from call to call. It is up to you to check this vehicle just as you would check your own car.

6. Write clear, detailed service reports for each job. This is necessary because someone in your service department will have to make up a bill to send to the customer for the work you have done. Even on jobs that have service contracts, this is important so a service history can be developed for each customer.

7. Be careful! Your safety and the safety of the people living in the buildings you work in are your responsibility. Remember that you are working with fire! Under normal operating conditions, parts of the equipment, particularly the flue pipe connection to the chimney, will get very hot. You have to avoid touching hot surfaces and make sure that there are no combustible materials (newspapers, cardboard boxes, clothing, and so forth) stored close to the heating equipment. Always check the oil piping connections at the burner. Oil leaks are not only smelly and messy, they present a serious fire hazard as well. Pay particular attention to the ignition system because the worst thing that can happen to an oil burner is a small but dirty explosion called a "puff back." There are no short cuts! If you follow the procedures explained in this book, you will become one of the people who can properly service and maintain oil-fired heating equipment so that the systems operate safely and efficiently. Treat every service call as if you were working in your own home.

The industry is looking for good people. Working conditions, salaries, and benefits will vary with geographic location. There is no doubt that a well-trained, competent oil burner service technician is a valuable member of any community regardless of where it is located. Another consideration in choosing this as a career is that the work is challenging and never boring. You never know what each service call will require of you. It is not like working on a factory assembly line.

This industry is not very old. Central heating systems were not installed in many buildings until around 1920. Most apartment buildings in major cities had "cold water flats." The heating system in some family homes was the kitchen range and possibly a fireplace or wood-burning stove someplace in the house. Only the most expensive residences had central heating systems. The type of system installed varied with geographic location. Steam systems were installed in areas where there were people experienced in steam piping because there were local industries that used steam. Hot water systems evolved in an attempt to correct problems with early steam systems. Warm air systems evolved from the wood- or coal-burning stove in the kitchen that attempted to heat the house. The installation of early warm air furnaces, ducts, and room registers marked the beginning of the HVAC industry that separated plumbing from this new mechanical trade. The oil burner did not emerge until almost 1930.

The early oil burners were terrible machines. They were big and clumsy and noisy. Each manufacturer made his burner so unique that none of the parts would fit any other machine. The only advantage they offered was that coal did not have to be shoveled and ashes removed. There was a great surge of oil burner installations when World War II ended in the late 1940s. Coal-fired boilers were converted to oil and oil-fired boilers were installed. Many new fuel oil companies were formed and jobs for service technicians opened up. There were some training schools for service of oil burners but most mechanics learned on the job. With little or no formal training it was inevitable that mistakes would be made. Customers drifted from company to company searching for someone who could keep their equipment operating properly. In disgust, many of them, often the best accounts, switched to alternative fuels.

The industry grew rapidly in the 1950s when large cities passed laws requiring central heating systems in multiple dwellings and entire communities of single family homes sprouted up all over the country. Fuel oil dealers actively pursued heating system installations as a method of securing accounts. The same problems persisted: poor service, untrained service people, and burners operating inefficiently. With the energy crunch of the 1970s, we suddenly got the message that for the survival of the industry we had better make these machines operate properly. The machinery has been improved and now with the rising cost of energy it is essential to train technicians to care for the equipment properly.

Oil burners operate as an integral part of one of three types of heating systems. It is necessary to understand the operation of the heating system as well as the operation of the oil burner itself. Your area may be dominated by one type of heating system but it is certainly to your advantage to become familiar with all of them. Listed below is a brief description of each system and how it operates. A separate chapter in this book is dedicated to a detailed description of each system.

STEAM HEATING SYSTEMS

You can recognize steam systems by the gauge glass—a Pyrex glass tube mounted between two valves on the side of the boiler near the top. This device allows you to see how much water is in the boiler. There is also a pressure gauge that reads from 0 to 15 psi and a steam safety valve. When there is a call for heat, the burner starts and heats the water in the boiler until it starts to generate steam. The steam fills the piping and enters the radiators in the building. The radiators heat the air in the rooms, which will satisfy the thermostat and shut off the burner. All of the steam in the system will then condense and return to the boiler where it will be ready for the next heating cycle. Electrical controls are wired in a circuit that will allow the system to operate safely and heat the building automatically.

HOT WATER HEATING SYSTEMS

The hot water systems were developed as an alternative to the steam systems. There are several variations of these systems but they all have a therm-altimeter gauge that has three separate scales on its face. The gauge indicates the temperature of the water in the system, the water pressure, and the altitude. The functions of the therm-altimeter gauge are discussed in Chapter 9, Hot Water Heating Systems.

There are two basic types of hot water systems—gravity and forced. These systems are completely filled with water. The boiler, piping, and radiation must be bled to remove all the air. The gravity type operates on the principle that heating water will make it lighter. In a system full of water, the lighter water will rise to the top. When the thermostat calls for heat, the burner starts and heats the water in the boiler. As the water is heated, it rises into the supply piping and eventually into the radiators. The colder water in the radiators slowly moves back to the boiler through the return piping. The hot water in the radiators warms up the air in the rooms and satisfies the thermostat that will turn off the burner. Heated water will continue to rise from the boiler to the radiators, which results in slow, even heating of the building, with the burner starting only when the thermostat calls for heat. This type of system must be on all the time because it takes a long time to get the house up to the desired temperature if the system is shut down for an extended period of time.

The forced system utilizes a circulating pump to push the hot boiler water into the radiators when the thermostat calls for heat. A hot water control is installed in the boiler to turn the oil burner on when the water temperature in the boiler drops. Electrical controls are wired in a circuit that will allow the system to operate safely and automatically.

Radiant Heating Systems

A variation of the force hot water heating system is the radiant heating system. These are usually found in new homes but can be retrofitted into existing buildings as well. The system consists of a source of hot water, either a standard boiler or a hot water heater, and a piping circuit embedded in a concrete slab, or in the outside walls, or in the ceiling, or attached to the underside of the floor. Warm water, usually not hotter than 105°F, is circulated through the piping loop that heats the entire surface of the floor, wall, or ceiling. Heat radiating from the surface will effectively warm the people in the heated space. A detailed description of the piping layout and controls can be found in Chapter 9, Hot Water Heating Systems.

WARM AIR HEATING SYSTEMS

Warm air heating systems are easily recognized because there is no boiler and connecting pipe work. These systems utilize a furnace that is connected to sheet metal ducts that distribute the warm air to the building and return cool air from the building to the furnace. There are two basic types of warm air systems—gravity and forced.

The gravity system operates on the principal that heating air will cause it to rise. When the thermostat calls for heat, the oil burner starts and heats the air in the furnace. The heated air rises into the building through the supply ducts while the cooler air from the building moves to the furnace through the return ducts. The process continues until the thermostat is satisfied and the burner shuts off. Some circulation will continue because the air in the furnace is still hot. This may result in moderate overheating of the building.

The forced system uses a large fan, commonly referred to as a blower, to circulate heated air through the ducts. On a call for heat, the burner starts and heats the air in the furnace. When the temperature of the air in the furnace reaches approximately 120°F, the blower starts and sends the warm air into the supply ducts to the building. Cooler air returns to the furnace where it is heated. The process continues until the thermostat is satisfied and the burner shuts off. The blower will continue to operate as long as there is warm air in the furnace. When the air in the furnace cools off, the blower stops and the unit waits for the next heating cycle. Electrical controls are wired in a circuit that will allow the system to operate safely and automatically.

Space Heaters

Kerosene-fired space heaters have been developed to provide highly efficient, dependable alternatives to conventional heating systems. They have fully automatic forced draft vaporizing pot type burners built into an attractive, compact cabinet that can be installed in any room or space in a building. A detailed explanation of the burner can be found in Chapter 3, Oil Burners. The operation of the entire unit, the electronic safety and programming controls, venting system, and air circulation system can be found in Chapter 10, Warm Air Heating Systems.

This book includes explanations of the science involved in heating systems. These explanations, as well as the service procedures outlined, have been simplified so that you can understand the operation of the burner and the heating system and become an effective service technician. There is no reason for you not to continue your education, even while working in the industry, to include areas such as Refrigeration and Air Conditioning, Heating System Design, Industrial Oil Burners, Solar Energy Collection, Plumbing, Electrical Wiring, and Building Insulation. These are all related fields.

Start your tool box as soon as possible so that you are familiar with the feel of your tools when you start working on the equipment. The list of tools below is suggested.

Suggested List of Tools for Oil Burner Service Technicians

1. A Solid Tool Box
2. A Good Flashlight—Circuit Tester Type
3. Wrenches
    8″ Adjustable Open End
    Combination Box—Open End—3/8, 7/16, 1/2, 9/16, 5/8
    Pipe Wrenches—8″, 10″, 12″, 14″

4. Pliers
   8″ Linesman
   Channel Lock Water Pump
   Long Nose
   Diagonal Cutting
   Crimping Tool
5. Screwdrivers—Flat Blade
   8″ Plastic Handle—Cushion Grip
   6″ Plastic Handle—Cushion Grip
   4″ Plastic Handle
   Phillips #2 Tip—Plastic Handle
   Nut Driver Set—1/8, 3/16, 1/4, 5/16
6. Set of 6″ Allen Keys
7. Gauge Glass Cutter
8. Tubing Cutter
9. Flaring Tool
10. Vacuum Gauge
11. Pressure Gauge—0–300 psi
12. Pocket Draft Gauge
13. Hacksaw
14. Extension Light
15. Hammer
16. Chisel
17. Pocket Knife
18. Nozzle Extractor
19. Flame Mirror
20. Test Light—120 Volt
21. Multimeter
22. High Voltage Tester for Ignition Transformers

To Briefly Summarize

1. This text is designed to assist in the training of oil burner service technicians.
2. Included are explanations of the systems and parts involved in oil burner work.
3. If you have the qualities required, you can have a rewarding and challenging career as an oil burner service technician.

Please Answer These Questions

1. What are the qualities that a service technician has to develop to be successful?
2. How does a steam heating system operate?
3. How does a hot water heating system operate?
4. How does a warm air heating system operate?
5. Please describe the heating system in your home.

CHAPTER

2

THE COMBUSTION PROCESS

The function of a heating system is to replace the heat lost from a building to the outside air. The colder it is outside, the greater the amount of heat the system must put into the building. Most heating systems are designed to maintain an inside temperature of 70°F when the outside temperature is at the coldest for that particular geographic area. For example, the heating system in a house in LaCrosse, Wisconsin would have to be larger than the heating system in the identical house in Atlanta, Georgia. Additional factors in determining the size of the heating system are the size and shape of the building; construction materials used; numbers, sizes, and types of windows; numbers, sizes, and types of doors; and orientation of the building to the sun.

FUEL

The source of heat for any heating system is called fuel. A fuel is a material that releases heat energy when it burns. To qualify as a fuel, a material must have these qualities:

1. There must be an abundant supply. It would not make any sense to develop a heating system based on a fuel that would be exhausted in a short period of time.
2. It must be relatively inexpensive.
3. It must be safe to transport and hold in storage until it is used.
4. It must have a high heat content that can be released easily and efficiently.
5. The heat content must be obtainable without polluting the air or causing environmental problems.
6. The material selected as a fuel should have no other value to the community.

The fuels we have been using are known as *fossil fuels*. They are the result of vegetation and animal life that was trapped under the surface of the earth during the prehistoric period in the formation of this planet. This process took millions of years to complete. The fuel that we are using cannot be replaced and will eventually be used up. The equipment described in this book is designed to release the heat energy in fuel oil.

FUEL OIL

Fuel oil is one of the products extracted from petroleum during the refining process. The petroleum, or crude oil, that we refine in this country comes from a variety of locations around the world. The petroleum from different locations will have some variations in impurities and consistency. These variations are corrected in the refining process so that uniform products are produced that can be used anyplace.

The refining process is basically distillation. Crude oil is heated and the resulting vapors rise into a condensing tower. The condensing tower contains a series of heated trays that the vapor passes over. Gasoline is inserted through the top of the tower to wash the condensing vapor onto the trays where another vaporization takes place. The vapor of the more volatile product will rise to the top of the tower where it condenses and is drained off into storage tanks. This would be gasoline. The fuel oils are collected in the middle range of the tower while the lubricating oils and residual fuel oil are the last of the products to be collected.

The Products of Refining Are:

Gasoline—used as our basic internal combustion engine fuel.

#1 Fuel Oil—Kerosene, which is used as fuel for jet aircraft and as domestic fuel in small buildings.

#2 Fuel Oil—Diesel oil, which is used for engines and is identified by the addition of a blue dye during the refining process.

#2 Fuel Oil—Is used as fuel in heating systems and is identified by the addition of a red dye during the refining process.

#4 Fuel Oil—The last product of the distillation process, before the lubricating and residual fuel oil, it is used as fuel in small commercial and apartment buildings. Generally cheaper than #2 oil, it does require more preparation for the combustion process than #2 oil.

#6 Fuel Oil—This is the residue that remains in the process when all other products have been extracted. It is used as fuel in large buildings and requires special equipment as well as preheating of the oil to burn properly.

The American Petroleum Institute has developed a system of identifying the various grades of fuel oil. The system known as the API Degrees of Gravity is based on the weight of specific petroleum products as compared to an equal volume of water. The API degree range for #2 fuel oil is from 30 to 38 degrees. This range clearly indicates that there are variations in #2 fuel oil that have to be dealt with when servicing oil-fired equipment. Some of the factors that have to be considered are:

1. The **VISCOSITY** of the oil increases as the API degrees decrease. The viscosity of a liquid describes how well it will flow. Think of two common liquids—water and honey. Water flows quite readily while honey does not. This is because honey has a

higher viscosity than water. Variations in viscosity, particularly in colder areas, may require the use of larger size oil line piping.

2. The **FLASH POINT** of the oil increases as the API degrees decrease. The flash point is the temperature at which the vapors escaping from the surface of the oil can momentarily be ignited by an external flame. The minimum flash point for #2 fuel oil is 100°F.

3. The **FIRE POINT** of the oil is approximately 20 degrees higher than the flash point. The fire point is the temperature at which the vapor escaping from the oil will ignite and continue to burn. Both the flash and fire point temperatures must be considered in the location of the fuel oil storage tank. Local building and fire codes must be adhered to in all installations of storage tanks and heating equipment.

4. The **POUR POINT** of the oil decreases as the API degrees increase. The pour point is the temperature at which the oil will congeal or turn into a solid. Oil does not freeze but wax crystals do form as the temperature drops. These wax crystals combine and can easily clog oil lines and filters. The pour point for #2 fuel oil is approximately 20°F. When #2 oil must be stored outside or in cold locations, it is advisable to switch to #1 fuel oil (kerosene) because it has a pour point of approximately 0°F.

5. The **HEAT CONTENT** of the oil increases as the API degrees decrease. The accepted heat value for one gallon of #2 fuel oil is 140,000 Btus. One Btu is the amount of heat required to raise the temperature of one pound of water one degree Fahrenheit. The heat content of #2 oil does vary from a low of 137,000 Btus to a high of 141,800 Btus depending on the weight of the oil. The heavier grades of oil, as indicated by lower API degree numbers, have a higher heat content per gallon.

6. **FRACTIONS.** #2 Fuel oil is one of the by-products of the crude oil distillation process. The vaporizing temperature for the portion of the crude oil that will condense and become #2 oil is in the range of 400 to 640°F. Because of this, the oil contains fractions that vaporize at different temperatures. The lightest fraction, which should be at least 10% of the total volume, will vaporize at 400°F. This is essential for the smooth ignition of the oil as it leaves the oil burner because the heat generated by the ignition spark is approximately 450°F. As the air–oil mixture passes by the ignition spark, the most volatile fraction of the oil will vaporize and ignite. The heat generated by the ignition of the volatile fraction will help to vaporize the rest of the oil and establish the flame. This all takes place in an instant. Problems may arise with the heaviest fraction in the oil depending on the API degrees of the particular sample of oil. If the API degrees of the sample are close to the minimum of 30, there will be a higher percentage of heavy fractions that requires more heat to vaporize, which might have an effect on the completion of the combustion process.

The values on Table 2-1 are approximate. The table should help you to understand the API degree system. The API degree number for a specific load of oil can be found on the shipping documents sent with the oil from the refinery to the bulk storage plant.

GRADE OF OIL	#1	#2	#4	#6
Degrees API				
min	38	30	20	8
max	45	38	28	15
Kinematic Viscosity				
cSt/s—100°F				
min	1.4	2.0	5.8	—
max	2.2	3.6	26.4	—
Saybolt Universal Viscosity				
SUS—100°F				
min	—	32.6	45.0	900
max		37.9	125.0	900+
Flash Point	100°F	100°F	130°F	140°F
Pour Point	0°F	20°F	20°F	30°F
Weight lb/gallon				
min	6.675	6.960	7.396	8.053
max	6.95	7.296	7.787	8.448
Btu/gallon				
min	132,900	137,000	143,000	151,300
max	137,000	141,800	148,100	155,900

Table 2-1 **Characteristics of Fuel Oils.**

THE COMBUSTION PROCESS

The chemical composition of all fuel oil is 85% carbon and 15% hydrogen. Carbon is an element that combines readily with a variety of other elements to form many of the most familiar materials we use. Coal, diamonds, graphite, and rubber are some of the items we are familiar with that contain carbon. In its natural form, it is a solid. In fuel oil, it has combined with a gas, hydrogen, to form a liquid. The most common form of hydrogen is as the element that combines with oxygen to form water. The impurities such as sulfur found in some crude oils can be removed during the refining process. There are many localities where this is a requirement of air pollution control laws.

The combustion process is the rapid oxidation of a material that releases heat and light. We are all familiar with another oxidation process that takes place when steel or iron is left exposed to air. The oxygen in the air combines with the iron to form ferrous oxide, commonly known as rust. The oxygen for the combustion process comes from the air we breathe. Air is only 21% oxygen. The remaining 79% is a gas called nitrogen. Nitrogen is an inert gas that does not combine readily with other elements. The nitrogen that goes through the combustion process with the oxygen just goes along for the ride and does not contribute to the process. Under certain adverse conditions nitrogen can be a cause of air pollution.

The chemical reaction of the combustion process can be expressed by the following simplified diagrams:

$$\text{Carbon} + \text{Air [Oxygen \& Nitrogen]} \rightarrow \text{Carbon Dioxide} + \text{Nitrogen} + \text{Heat \& Light}$$

$$C + [O_2 + N_2] \rightarrow CO_2 + N_2 + \text{Heat \& Light}$$

$$\text{Hydrogen} + \text{Air [Oxygen \& Nitrogen]} \rightarrow \text{Water} + \text{Nitrogen} + \text{Heat \& Light}$$

$$2H_2 + [O_2 + N_2] \rightarrow 2H_2O + N_2 + \text{Heat \& Light}$$

As you can see, the carbon in the oil combines with oxygen to form carbon dioxide. The hydrogen in the oil combines with oxygen to form water. The nitrogen in the air just goes through this process without combining with anything. All the nitrogen does is get hot. The carbon dioxide, water, and nitrogen form what we refer to as flue gas. It is easy to see that we need a lot of air to supply the oxygen for the combustion process. In a laboratory under carefully controlled conditions, the perfect combustion process requires 14.36 pounds of air for every pound of fuel oil that is burned (see Table 2-2). Another way of saying this is that for every gallon of fuel oil we burn we must supply 1,540 cubic feet of air to ensure that we have the correct amount of oxygen for the combustion process. We can check the combustion process by determining the percentage of carbon dioxide in the flue gas. The perfect combustion process will result in 15.3% carbon dioxide in the flue gas.

As a service technician you are in control of the combustion process. You can control the amount of air entering the flame. You can also control the amount of oil entering the flame. By either reducing the air or increasing only the oil, the chemical reaction will change to something like this:

$$(C+H) + (O+N) \rightarrow CO_2 + CO + H_2O + N + C + \text{HEAT} + \text{LIGHT}$$

The carbon monoxide (CO) produced will vary depending upon how incomplete the combustion process is in each particular unit. Obviously, this should be avoided since carbon monoxide is poisonous and even small quantities can adversely affect the health of the people in the building. The carbon (C) that results from this chemical reaction remains in the boiler, and all over the boiler room, as soot. Soot is carbon in the form of lightweight dust that floats all over the building. Soot has been one of the most serious problems faced by the fuel oil industry. This is where the problem starts. A service technician who knows nothing about the combustion process leaves a poorly adjusted oil burner, and there goes an unhappy customer switching to another oil company. The sign that soot is forming is a smoking boiler. Do not leave a job that is showing signs of smoke coming out of the chimney.

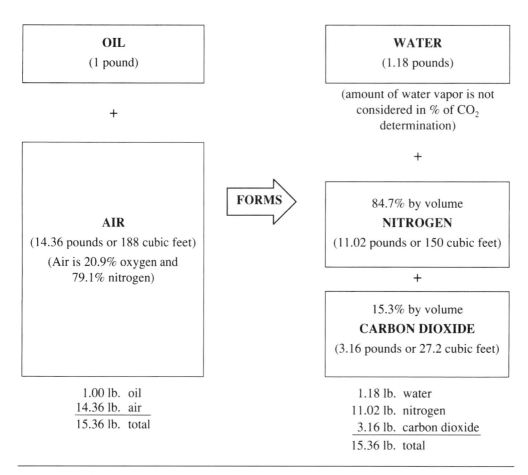

Table 2-2 **The Results of a Perfect Combustion Process That Is Possible in a Laboratory.** *Courtesy: R. W. Beckett Corp.*

The answer *seems* to be to just open the air shutter and give the flame more air than it needs. This is not the solution either because the result would be as shown on Table 2-3 or the diagram below:

$$(C+H) + (O+N) \rightarrow CO_2 + H_2O + N_2 + NO_x + O_2 + HEAT + LIGHT$$

Increasing the air changes the composition of the flue gas. Free oxygen is released that will attack the flue breeching and rust it out. This adjustment will not produce soot but it will produce nitrogen oxide (NO_x) which is an air pollutant with an annoying odor and that stings the eyes.

It is almost impossible to duplicate the results of the laboratory combustion process in the field. We have to make an attempt to get as close to those results as possible. Later in this book (Chapter 19) we will discuss testing and adjusting oil burners. It is absolutely

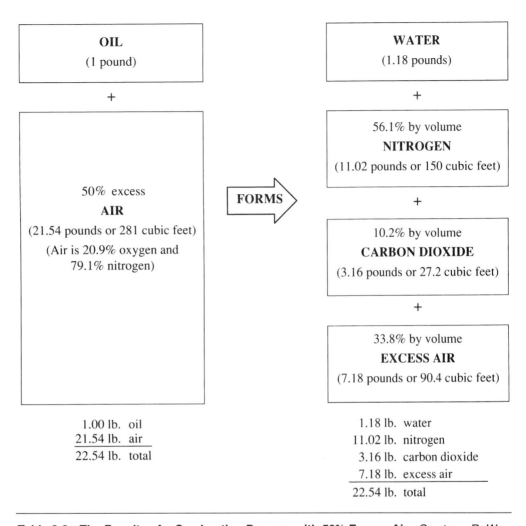

Table 2-3 **The Results of a Combustion Process with 50% Excess Air.** *Courtesy: R. W. Beckett Corp.*

necessary for you to completely understand the combustion process before you can begin to understand how oil burners operate.

COMBUSTION CHAMBERS

The combustion process works best if the flame burns in suspension, enclosed in a combustion chamber made of refractory materials that will reflect heat into the flame and help to completely vaporize the oil. A properly designed combustion chamber made of appropriate materials leads to complete combustion of the fuel, higher efficiency, and satisfied customers.

When the first oil burners were installed in boilers that were built to be fired by coal, the combustion chamber was used to protect the base of the boiler. Little or no consideration was made for the flame pattern or size as produced by the burner. This is no longer the case. When you purchase an oil burner the instructions sent with the machine contain a detailed description of how the combustion chamber should be built (see Figure 2-1 and Table 2-4). Table 2-5 contains dimensions of combustion chambers for various types of oil burners. The square inches per gallon relates to the floor area of the chamber. The 80 square inches per gallon section is for flame retention type burners because their flame patterns are very compact and the flame stays close to the burner end cone. The 90 square inches per gallon is for conventional high pressure, gun type burners because their flame patterns drift somewhat from the end cone and need additional space to burn. The 100 square inches per gallon section is for low pressure, air-atomizing burners because their larger flame patterns require the additional space to burn completely.

The flame should fit into the combustion chamber like your hand fits into a glove (see Figure 2-2 on page 18). There should be no contact with the floor or side walls of the chamber. Oil hitting the chamber walls or floor is called impingement and will lead to the deposit of carbon on the floor or walls that will continue to grow until the situation is corrected.

Chambers were made of hard fire brick. These bricks are not insulating and must be backed with some form of insulation. For many years the insulation used was asbestos. Extreme care must be taken if you have to remove an old hard brick combustion chamber. Check your local laws and regulations regarding the handling and disposal of asbestos

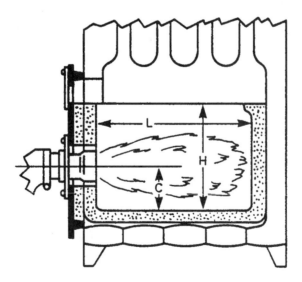

Figure 2-1 **Combustion chamber dimension diagram for Beckett AF oil burner.**
Courtesy: R. W. Beckett Corp.

1 FIRING RATE (GPH)	2 LENGTH (L)	3 WIDTH (W)	4 DIMENSION (C)	5 SUGGESTED HEIGHT (H)	6 MINIMUM DIA. VERTICAL CYL.
0.50	8	7	4	8	8
0.65	8	7	4.5	9	8
0.75	9	8	4.5	9	9
0.85	9	8	4.5	9	9
1.00	10	9	5	10	10
1.10	10	9	5	10	10
1.25	11	10	5	10	11
1.35	12	10	5	10	11
1.50	12	11	5.5	11	12
1.65	12	11	5.5	11	13
1.75	14	11	5.5	11	13
2.00	15	12	5.5	11	14
2.25	16	12	6	12	15
2.50	17	13	6	12	16
2.75	18	14	6	12	18

Table 2-4 Combustion Chamber Dimensions for Beckett AF Oil Burner. *Courtesy: R. W. Beckett Corp.*

					HEIGHT FROM NOZZLE TO FLOOR (INCHES)			
Oil Consumption GPH	Square Sq.Inch Area Combustion Chamber	Dia. Round Combustion Chamber Inches	Rectangular Combustion Chamber Inches	Combustion Chamber Inches	Conventional Burner W × L	Conventional Burner Single Nozzle	Sunflower Flame Burner Single Nozzle	Sunflower Flame Burner Twin Nozzle
80 Sq. In. Per Gal.								
.75	60	8 × 8	9	—	5	x	5	x
.85	68	8.5 × 8.5	9	—	5	x	5	x
1.00	80	9 × 9	10 1/8	—	5	x	5	x
1.25	100	10 × 10	11 1/4	—	5	x	5	x
1.35	108	10 1/2 × 10 1/2	11 3/4	—	5	x	5	x
1.50	120	11 × 11	12 3/8	10 × 12	5	x	6	x
1.65	132	11 1/2 × 11 1/2	13	10 × 13	5	x	6	x
2.00	160	12 5/8 × 12 5/8	14 1/4	6	x	7	x	x
2.50	200	14 1/4 × 14 1/4	16	12 × 16 1/2	6.5	x	7.5	x
3.00	240	15 1/2 × 15 1/2	17 1/2	13 × 18 1/2	7	5	8	6.5
90 Sq. In. Per Gal.								
3.50	315	17 3/4 × 17 3/4	20	15 × 21	7.5	6	8.5	7
4.00	360	19 × 19	21 1/2	16 × 22 1/2	8	6	9	7
4.50	405	20 × 20		17 × 23 1/2	8.5	6.5	9.5	7.5
5.00	450	21 1/4 × 21 1/4		18 × 25	9	6.5	10	8

100 Sq. In. Per Gal.

5.50	550	23 1/2 × 23 1/2	Round	20 × 27 1/2	9.5	7	10.5	8
6.00	600	24 1/2 × 24 1/2	combustion	21 × 28 1/2	10	7	11	8.5
6.50	650	25 1/2 × 25 1/2	chambers	22 × 29 1/2	10.5	7.5	11.5	9
7.00	700	26 1/2 × 26 1/2	usually	23 × 30 1/2	11	7.5	12	9.5
7.50	750	27 1/4 × 27 1/4	not used	24 × 31	11.5	7.5	12.5	10
8.00	800	28 1/4 × 28 1/4	in these	25 × 32	12	8	13	10
8.50	850	29 1/4 × 29 1/4	sizes	25 × 34	12.5	8.5	13.5	10.5
9.00	900	30 × 30		25 × 36	13	8.5	14	11
9.50	950	31 × 31		26 × 36 1/2	13.5	9	14.5	11.5
10.00	1000	31 3/4 × 31 3/4		26 × 38 1/2	14	9	15	12
11.00	1100	33 1/4 × 33 1/4		28 × 29 1/2	14.5	9.5	15.5	12.5
12.00	1200	34 1/2 × 34 1/2		28 × 43	15	10	16	13
13.00	1300	36 × 36		29 × 45	15.5	10.5	16.5	14
14.00	1400	37 1/2 × 37 1/2		31 × 45	16	11	17	14.5
15.00	1500	38 3/4 × 38 3/4		32 × 47	16.5	11.5	17.5	15
16.00	1600	40 × 40		33 × 48 1/2	17	12	18	15
17.00	1700	41 1/4 × 41 1/4		34 × 50	17.5	12.5	18.5	15.5
18.00	1800	42 1/2 × 42 1/2		35 × 51 1/2	18	13	19	16

Table 2-5 **Combustion Chamber Dimensions for Various Types of Oil Burners.** *Courtesy: R. W. Beckett Corp.*

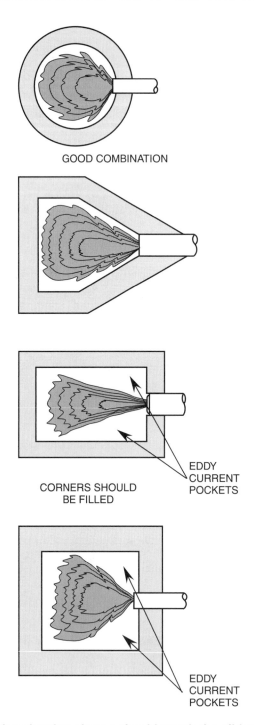

GOOD COMBINATION

CORNERS SHOULD
BE FILLED

EDDY
CURRENT
POCKETS

EDDY
CURRENT
POCKETS

Figure 2-2 **Combustion chamber shapes should match the oil burner flame for best results.** *Courtesy: R. W. Beckett Corp.*

before you start removing the chamber. The use of disposable safety suits, gloves, and respirators is highly recommended.

It takes a long time for hard fire brick to get hot enough to help in the vaporization of the oil. For this reason, they are not really suited for small oil burners that have short running cycles. Soft, insulating firebrick is best for small burners because the heat does not pass through the brick, which allows the inside surface to become red hot quickly. This heat radiates back into the flame, helping to vaporize the oil completely, which has a positive impact on the combustion process. There are some new materials—products of the space program—that are popular now, such as cerafelt and ceramic fiber combustion chambers. There is a product called a *wet pack* that is like a blanket that can be installed inside an existing chamber to improve the operation of the oil burner.

Once the material is selected, the chamber must be built to contain the flame. The chamber size is based on the gallons per hour (GPH) that the oil burner will be burning. For every gallon, up to 3 gallons per hour, you will need 80 to 90 square inches of floor area (see Figure 2-3). The distance from the center line of the nozzle to the floor (x) is

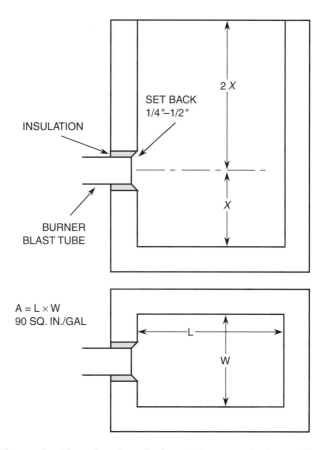

Figure 2-3 Basic combustion chamber design—90 square inches of floor area for every gallon per hour firing rate.

critical. Check the burner manufacturer's specifications if they are available. If they are not, the minimum distance for a small oil burner should be 6 inches. The height from the center line of the nozzle should be twice the distance to the floor. This should protect the base of the boiler, if it has one, and supply the proper environment for the combustion process. The shape of the chamber will be determined by the flame pattern of the oil burner. Different burners produce flames of differing shapes (see Figure 2-4). The flame has to fit without hitting the surface of the chamber. This does not mean that you can make it just a little bigger and be safe. Try to avoid dead air spaces in the chamber because they will reduce the efficiency of the burner. Experience and following manufacturer's specifications will help you to build the combustion chamber that is right for each burner you install.

Many new wet base boilers are equipped with flame retention head oil burners that can operate efficiently with no combustion chamber or a partial chamber. This can be a refractory target wall or floor that will reflect heat into the flame for complete combustion. Precast, hard brick combustion chambers are available for most boilers (see Figure 2-5). They are approximately one inch thick, which allows them to transfer heat to the boiler as well as to heat up quickly, radiating heat back into the flame. The use of a combustion chamber, even in a new boiler that can operate without one, will improve the operating efficiency due to the improvement in the combustion process.

Designing combustion chambers to improve combustion efficiency takes experience and some experimentation. The use of wing walls at the front of the chamber will help to vaporize the oil as it leaves the nozzle (see Figure 2-6). This is a most critical area of the chamber since this is where the air and oil must mix for complete combustion. The heat radiating from the wing walls will convert the tiny droplets of oil leaving the nozzle into

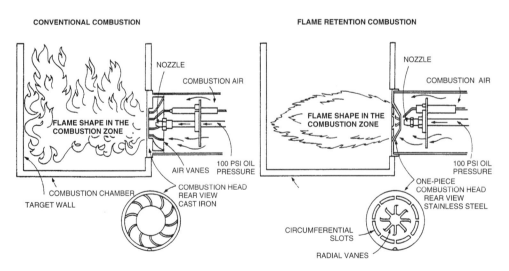

Figure 2-4 **Flame shape of conventional and flame retention head oil burners.** *Courtesy: R. W. Beckett Corp.*

Figure 2-5 **Precast combustion chambers are designed to fit specific boilers.**

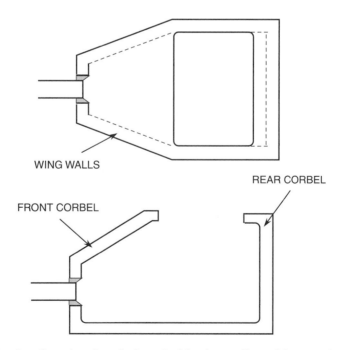

Figure 2-6 **Combustion chamber designed with wing walls and front and rear corbelling. This type of chamber reflects heat directly into the flame for improved vaporization of the oil.**

a vapor that can readily mix with the air and burn completely. Another effective design feature is corbelling (see Figure 2-7). A corbel is made by laying refractory material across the top of the chamber at the rear or front or both places. The idea is to reflect heat into the flame from the top that will vaporize the heaviest fractions in the oil and ensure that no oil passes through the combustion chamber without burning completely. Care must be taken not to restrict the movement of the large volume of gas released by the combustion process. This could lead to smoke and the buildup of soot in the boiler. Another concern of "over corbelling" a combustion chamber is the heat that will remain at the end of the firing cycle. Excessive heat due to the restrictions in the boiler flue gas passages, stack dampers, or poor draft conditions will radiate onto the nozzle causing the oil left

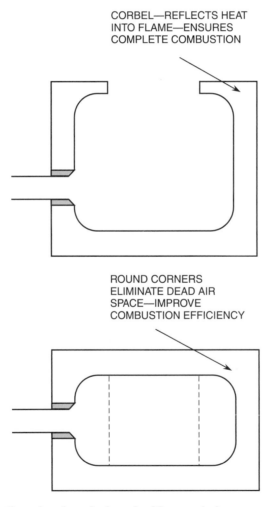

Figure 2-7 **Combustion chamber designed with rounded corners and corbels for improved fuel vaporization.**

at the end of the firing cycle to drip out of the nozzle. This will create a potential oil leak problem and cause carbonization of the nozzle exterior.

To Briefly Summarize

1. There are variations in fuel oil as expressed in the API degree system. These variations will affect the operation of oil-fired equipment. Service technicians must be aware of the properties of the fuel in their particular geographic area.
2. The combustion process is a chemical reaction during which heat energy stored in the fuel is released.
3. As a service technician, you control the combustion process through your ability to change the air and fuel adjustments of the oil burner.
4. The combustion process works best in the environment of a combustion chamber that reflects heat into the flame to help vaporize the fuel oil.

Please Answer These Questions

1. What is the significance of the API degree system?
2. How can service technicians use the information supplied by the API degree system?
3. What is the chemical composition of fuel oil?
4. What is the chemical composition of air?
5. As a service technician, you are in control of the combustion process. Explain why that statement is true.
6. What effect, if any, would a smoking oil burner have on the operation of the heating system?
7. What does *impingement* mean and why is it a problem?
8. You are sent to replace a combustion chamber. List some of the things you will have to do to install the best chamber for that particular oil burner.
9. Explain how you can improve combustion by modifying the design of a combustion chamber.
10. How can "over corbelling" a combustion chamber create problems with the oil burner?

CHAPTER 3

OIL BURNERS

An oil burner is a machine composed of component parts designed to prepare liquid fuel for combustion, start the combustion process, and project the resulting flame into a boiler or furnace in which the heat energy released from the fuel can be utilized. In order to accomplish this task, the machine must be able to convert the liquid fuel into a vapor that can be mixed with air so the fuel burns completely. Oil burners are classified by the method used to prepare the fuel for combustion.

VAPORIZING OIL BURNERS

Vaporizing oil burners were among the earliest machines used for domestic heating purposes. They were designed to burn kerosene because it is more volatile than #2 oil and therefore, easier to vaporize. There are two types of vaporizing oil burners. They are the forced draft and the natural draft burners. The forced draft burners were used in regular boilers or furnaces and operated automatically. They are now obsolete and have been replaced by the high pressure gun type oil burner. The natural draft burners are used in cabinet type space heaters in small buildings that do not require a central heating system. They have the advantage of being able to burn at a very low gallons-per-hour rate that is almost impossible to achieve with a higher pressure gun type oil burner. However, they are not automatic and require supervision by someone who knows how to operate them as a safe source of heat.

The fuel delivery system requires that the oil tank be installed above the level of the burner pot (see Figure 3-1). This allows gravity flow to the float operated constant level valve that is mounted on the outside of the cabinet. A metering valve is attached to the constant level valve. This device controls the flow of kerosene to the burner pot and is used to regulate the size of the flame.

The bottom of the pot is solid and has a connecting port for the oil line from the metering valve (see Figure 3-2). The side walls of the pot are perforated to allow combustion air to be drawn into the pot where it can mix with the kerosene vapors. There are rings inside the sleeve where the vapor-air mixture burns. As these rings heat up, they in-

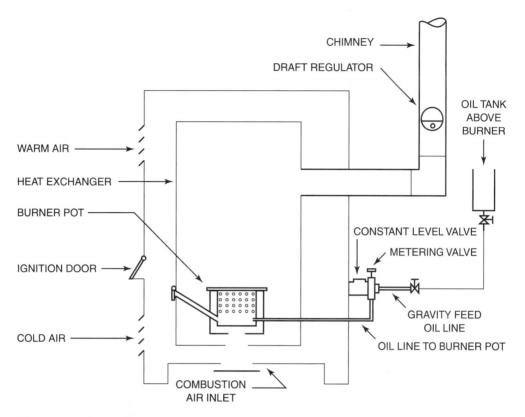

Figure 3-1   **Space heater with natural draft vaporizing pot type burner.**

crease the rate of vaporization and aid in complete combustion of the fuel. Some burners have a spreader head where the flame comes out of the pot. This head forces the shape of the flame to look something like a sunflower as the vapors burn at the edge of the pot. When the spreader head gets hot, it will ensure that no fuel vapors leave the burner without burning completely.

The Basic Operation of the Burner Is as Follows:

1. The metering valve is opened to the pilot setting. A small amount of fuel runs into the bottom of the pot.
2. An external source of heat, usually a long match, is inserted through the lighting door and held above the surface of the kerosene in the pot. The heat supplied from the match will vaporize some of the kerosene and a flame will be established.
3. The metering valve is opened slowly to the next setting. This allows more kerosene to flow into the pot and that increases the size of the flame. As the flame gets larger, more air is drawn into it through the perforations in the side walls of the pot.
4. The lowest flame ring on the side wall heats up and a steady flame is established.

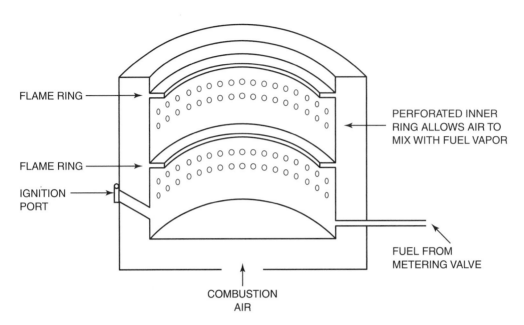

FLAME RING

PERFORATED INNER
RING ALLOWS AIR TO
MIX WITH FUEL VAPOR

FLAME RING

IGNITION
PORT

FUEL FROM
METERING VALVE

COMBUSTION
AIR

Figure 3-2   **A natural draft vaporizing pot.**

5. The metering valve can now be opened to increase the size of the flame according
   to the heating needs of the building.

   These burners require the use of the highest quality fuel available. Fuel line filters
and strainers must be cleaned or replaced on a regular basis. The air intake port and the
side wall perforations must be kept clean. If carbon deposits build up inside the side wall
sleeves, the air will not be able to mix with the fuel vapors, causing the formation of
smoke and soot.

   Another serious problem with these units is that they rely on the air in the space
being heated for the combustion process. In most installations, there is enough air infil-
trating through spaces around doors and windows to keep the burners supplied with the
necessary air. Air comes into the space every time the outside door opens as well. However, in a tightly sealed building, a burner could use all the available air over a short period
of time. This would result in incomplete combustion of the kerosene, causing smoke,
soot, the formation of carbon monoxide, and potential health hazards to people in the
space being heated. The person in charge of the heater must be made aware of this danger. To avoid problems relating to insufficient air, a window must be left cracked open
regardless of the outside temperature.

   The latest improved versions of the vaporizing pot type burner can be found in
space heaters developed in Japan. They are forced draft burners that use a wall mounted
double pipe unit to bring combustion air into the burner and allow the exhaust gases to
leave the building (see Figure 3-3). The burner pot is at the bottom of the sealed com-

Figure 3-3 **Toyostove Laser 73 forced draft kerosene space heater.** *Courtesy: Toyotomi, U.S.A. Inc.*

bustion chamber. The fuel storage tank must be installed above the level of the burner. Fuel is delivered to the bottom of the pot through a float operated constant level valve and a four-position electrically operated metering device. The pot is heated by a ceramic ignitor that fits in a shielded bracket inside the pot. Combustion air is forced into the pot through an air channel connected to the air intake fan. On a call for heat, the air intake fan starts and purges the combustion chamber of any residual fumes or vapor. Once this is done and an air flow established, the ignitor is turned on and the pot heats up. A small amount of fuel is allowed to flow into the pot where it vaporizes, mixes with the combustion air, and is ignited. A flame rod senses the flame and allows the burner to continue to operate. The fuel metering device will then be able to control the flow of fuel automatically in response to the settings on the space heater and the temperature requirements of the space being heated. The flue gas is forced through the heat exchanger and then through the inner tube of the double pipe wall unit to the outside of the building (see Figure 3-4). A separate circulation fan forces the room air through the heat exchanger to deliver heat into the space. These units are controlled by an electronic microprocessor and are fully programmable and automatic (see Figure 3-5). Additional information on the operation of space heaters can be found in Chapter 10, Warm Air Heating Systems.

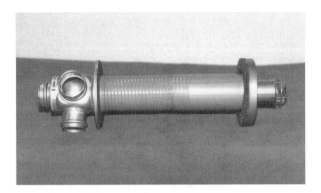

Figure 3-4 **Wall mounted double pipe for air intake and flue gas exhaust.** *Courtesy: Toyotomi U.S.A., Inc.*

Figure 3-5 **Internal view of Laser 73. Burner pot is in the lower left corner, the combustion chamber is directly above the pot, the heat exchanger is in the center, the control panel is on the right side, and the constant level valve and fuel solenoid assembly are in the lower right corner.** *Courtesy: Toyotomi U.S.A., Inc.*

VERTICAL ROTARY OIL BURNERS

Vertical rotary oil burners were popular as replacements for coal in round boilers as well as other applications (see Figure 3-6). A hearth made of castable refractory material is built at the top of the boiler base. The burner is set below the hearth so that the motor shaft extends vertically through the center of the hearth. On the motor shaft is a fan and a distributor head. Along the outer edges of the hearth, right next to the boiler, a track and metal grid are installed. Two large electrodes are installed in the hearth so that a high voltage spark will jump from the tip of the electrode to the track to ignite the air–oil mixture delivered by the rotating distribution head.

The fuel delivery system requires that the oil tank be installed above the level of the oil burner. This allows gravity flow to the constant level valve that is installed next to the boiler at a specific height above the top of the oil distributor. The constant level valve assembly includes a metering device and a solenoid valve to control the flow of oil to the distributor head.

The Basic Operation of the Burner Is as Follows:

1. When the burner starts, the fan establishes an air stream along the surface of the hearth.
2. The ignition system provides a spark at two points along the grid.

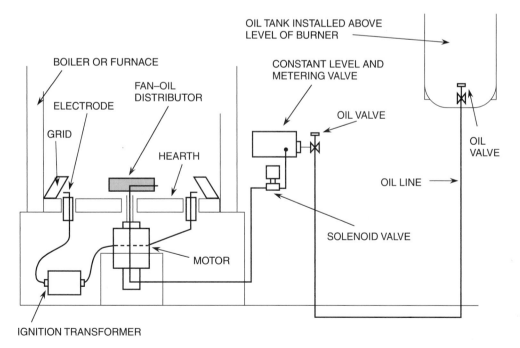

Figure 3-6 **Vertical rotary oil burner.**

3. The solenoid valve opens, allowing oil to flow to the distributor tubes.
4. The oil leaving the distributor tubes is atomized by the centrifugal force created by the rotation of the motor shaft.
5. The atomized oil mixes with the air from the fan and this mixture is ignited by the spark along the grid.
6. Once the flame is established, the metal grid gets hot, which helps to vaporize the oil for complete combustion.

These burners are tricky to adjust so the use of test instruments (see Chapter 18) is highly recommended. The final adjustment must be made with the boiler or furnace door closed because if it is open, excess air will be drawn in that will affect the flame. The ideal flame would be one that burns approximately 1/4″ above the grid, with a 2″ blue section that leads into a bright yellow flame 4–6″ high.

HORIZONTAL ROTARY OIL BURNERS

Horizontal rotary oil burners can be used to burn heavier grades of fuel oil (see Figure 3-7). The oil burner consists of a tapered brass cup that is mounted on a hollow shaft. Inside the shaft is the fuel tube that extends into the rear of the cup. On the shaft behind the cup, is a flat disc-type fan. The shaft assembly is encased in a housing that has an air intake opening that leads into the back of the fan. There is an air adjustment shutter or damper to control the amount of air entering the burner. The burner housing contains a pressure plate that fits directly in front of the fan. Inside the pressure plate is a series of vanes that directs the air from the fan into an air nozzle that fits over the outside of the tapered brass cup. There are vanes inside the air nozzle that force the air from the fan to leave the air nozzle at an angle.

The Basic Operation of the Burner Is as Follows:

1. When the burner starts, the shaft turns the fan and the cup at 3,450 rpm.
2. This establishes a high velocity air stream around the open end of the cup.
3. An external ignition system composed of a spark-ignited gas flame, establishes a gas pilot flame in front of the cup.
4. When the flame safety system confirms the presence of the gas pilot, the oil solenoid valve opens and fuel is delivered to the rear end of the spinning cup through the fuel tube.
5. Centrifugal force causes the oil to be thrown off the end of the cup in a fine film. The high velocity air stream breaks into the film and atomizes the oil.
6. The resulting air–oil mixture is ignited by the gas pilot and the flame is established.
7. The gas pilot will shut off once the flame safety system confirms the oil flame.
8. On shutdown the oil solenoid valve will close, stopping the flow of oil to the cup.
9. The burner will continue to operate for approximately 30 seconds to purge the combustion chamber of flue gas.

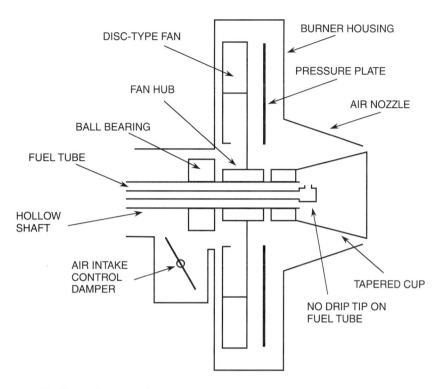

Figure 3-7   **Horizontal rotary oil burner.**

Preheating the oil lowers the viscosity so that it will atomize as it leaves the cup. The fan cannot deliver the amount of air required for the complete combustion of the oil. An external source of air must be provided. This is accomplished by the installation of a wind box that contains a powerful fan to deliver the additional air. Particular attention must be paid to the construction of the combustion chamber. The burner manufacturer's specifications must be followed because the complete vaporization of the fuel can only be accomplished with the reflected heat supplied by the refractory materials in the chamber. Daily maintenance of the burner is highly recommended. The cup and air nozzle must be cleaned to remove carbon that will form on these parts. Failure to maintain a clean burner will cause the burner to smoke and could eventually lead to a fire.

LOW PRESSURE AIR ATOMIZING BURNERS

Low pressure air atomizing oil burners are built in a variety of sizes for domestic and industrial applications (see Figure 3-8). The domestic burners are designed to burn #2 fuel oil while the industrial burners can burn #4 or #6 fuel oil. They have become extremely

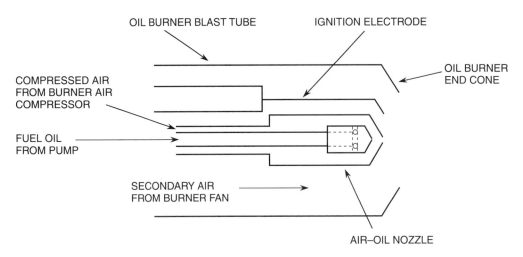

Figure 3-8   **Low pressure air atomizing oil burner.**

popular in larger sizes as replacements for the horizontal rotary burners. The burner assembly, regardless of size, must include an air pump or compressor and an air–oil nozzle.

The Basic Operation of the Burner Is as Follows:

1. When the burner starts, compressed air and fuel oil are forced into the air–oil nozzle. The compressed air is known as primary air.
2. The nozzle is designed to allow the air to break into the oil stream and mix with the oil.
3. As the air–oil mixture leaves the nozzle, the secondary air from the burner fan enters the mixture.
4. Direct electric ignition can be used on #2 and some #4 burners to ignite the air–oil mixture. A gas ignition system must be used for #6 oil burners.
5. When the burner shuts off, the compressed air in the line to the nozzle will blow through the nozzle and keep it clean.

Heavy grades of fuel oil must be preheated to reduce their viscosity so that the fuel will atomize. On larger installations, additional air may be required for complete combustion of the fuel. The manufacturer's specifications regarding combustion chamber design must be adhered to so that complete combustion of the oil is accomplished.

HIGH PRESSURE GUN TYPE OIL BURNERS

High pressure gun type oil burners are the most popular burners for domestic use (see Figure 3-9). They are designed to burn #2 fuel oil. These burners dominate the domestic oil burner field for several reasons:

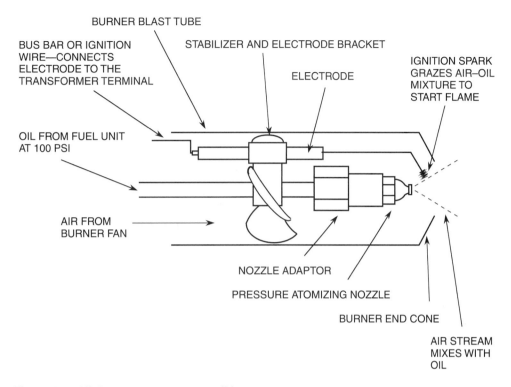

BURNER BLAST TUBE

BUS BAR OR IGNITION
WIRE—CONNECTS
ELECTRODE TO THE
TRANSFORMER TERMINAL

STABILIZER AND ELECTRODE BRACKET

ELECTRODE

IGNITION SPARK
GRAZES AIR–OIL
MIXTURE TO
START FLAME

OIL FROM FUEL UNIT
AT 100 PSI

AIR FROM
BURNER FAN

NOZZLE ADAPTOR

PRESSURE ATOMIZING NOZZLE

BURNER END CONE

AIR STREAM
MIXES WITH
OIL

Figure 3-9 **High pressure gun type oil burner.**

1. Regardless of which company builds the burner they all operate the same way.
2. There are few specialized parts and most of the parts are readily available.
3. The research and development investment by the burner manufacturers has resulted
   in excellent equipment that operates at high efficiency.

This book will deal primarily with the high pressure gun type oil burner. Most of the material is applicable to other burners as well.

Every oil burner must contain three separate systems: the air delivery system, the fuel delivery system, and the ignition system. There are external parts to these systems as well, but for now we will discuss only the parts that are built into the machine. Please understand that all the parts of these machines have specific functions. There are no decorative pieces. When you work on an oil burner there can be no parts left over. They all belong on the machine.

The largest single part of the oil burner is the main housing or chassis. This is a carefully designed casting that all the other parts attach to or fit inside of. The shape, which is called a *scroll*, is designed for the smooth movement of air. The housing has a large opening on one side (see Figure 3-10) and a smaller opening on the opposite side (see Figure 3-11). The larger opening is for the motor and the smaller opening on the opposite side is for the fuel pump. These openings are designed so that when the motor

Figure 3-10 **Beckett AFG oil burner chassis showing the motor mounting opening.**
Courtesy: R. W. Beckett Corp.

Figure 3-11 **Carlin EZ-1 chassis showing air intake slots and pump mounting opening.**
Courtesy: Carlin Combustion Technology, Inc.

and fuel pump are installed, their shafts line up perfectly. There are additional openings in the chassis in the area of the fuel pump. These are a series of slots that allow air to enter the burner. There is some type of band or shutter that covers these slots. This is how you adjust the amount of air that is used in the combustion process. The housing ends with a long tube, called the *blast tube*, which extends up to the inside edge of the combustion chamber of the boiler (see Figure 3-12).

The motors used for oil burners are generally split-phase motors or split capacitor motors. We will discuss these motors in some detail in Chapter 12, Electrical Equipment. New oil burners use motors that run at 3,450 rpm while older burners operate at 1,725 rpm (see Figure 3-13). Mounted on the motor shaft is the fan, sometimes referred to as the *blower wheel* (see Figure 3-14). The fan is made up of a series of blades attached to a back plate and held together by a front ring. These fans are balanced at the factory by placing small clips on one or two blades. Do not ever remove these clips because if you do the fan will vibrate and eventually fall apart. At the end of the motor shaft is the oil pump coupling. This unit has metal or plastic ends with a rubber shaft between them. One end fits on the motor shaft. The other end fits on the fuel pump shaft. The motor and pump have different size shafts. There can be no mistake made when installing the pump coupling since it only fits one way.

Figure 3-12   Beckett AFG oil burner chassis with the blast tube attached. *Courtesy: R. W. Beckett Corp.*

Figure 3-13   Motor, fan, and oil burner coupling for the Carlin EZ-1 oil burner. *Courtesy: Carlin Combustion Technology, Inc.*

Figure 3-14   Oil burner fan (blower wheel).

To briefly recap what was just described: we have an oil burner chassis with the motor attached; the fan and oil pump coupling are on the motor shaft; the fuel pump is mounted opposite the motor and the pump coupling is on the pump shaft; and there is a band around the air intake slots to control the amount of air entering the combustion process.

Another major part is the ignition transformer (see Figure 3-15). This device converts the 120-volt power supply to 10,000 volts, which is hot enough to ignite the air–oil mixture as it leaves the oil burner. Ignition transformers are located on the top or side of the chassis. The high voltage terminals are connected to the electrodes by either bus bars or high tension wire. The electrodes are attached to a bracket that fits on the burner drawer assembly (see Figure 3-16). This assembly also includes the oil burner nozzle and it fits

Figure 3-15 **Ignition transformer, with cad cell, for Beckett AFG oil burner.** *Courtesy: R. W. Beckett Corp.*

Figure 3-16 **Drawer assembly that includes the nozzle, the electrodes, the bus bars, and the flame retention head, for a Wayne Blue Angel oil burner.** *Courtesy: Wayne Home Equipment.*

into the blast tube of the burner. It is designed to be easily removed from the burner for servicing. There are additional parts such as the end cone, the stabilizer, copper tubing from the pump to the drawer assembly, and possibly some kind of locking device that keeps the drawer assembly positioned properly (see Figure 3-17).

When All of the Parts Are in Place, This Is How the Burner Operates:

1. On a call for heat, power is sent to the oil burner motor and the ignition transformer.
2. The ignition transformer sends 10,000 volts through the bus bars to the electrodes, establishing an active high voltage spark at the electrode tips that are positioned above and slightly in front of the nozzle.
3. The motor starts and the fan forces air through the scroll into the blast tube, through the blast tube, and out the end of the burner into the combustion chamber.
4. As the motor runs, it turns the fuel pump, which sends fuel oil at 100 psi, through the copper tubing into the drawer assembly and then through the nozzle.
5. As the tiny drops of oil leave the nozzle, they are surrounded by air coming out of the blast tube. This is where the mixing of the air and oil takes place.
6. The air–oil mixture passes by the 10,000-volt spark, which is hot enough to ignite it and turn the flame on.
7. Once the flame is established, it is no longer necessary to keep the ignition system operating. The ignition shuts off in many burners at this point. There are an equal number, or possibly more, in which the ignition system is on during the entire running cycle. The choice is yours!
8. When the call for heat is satisfied, the power to the burner shuts off and the machine stops, waiting patiently for the next call for heat.

Figure 3-17 Cutaway view of an Aero SV forced draft oil burner shows position of drawer assembly. *Courtesy: Aero Environmental Ltd.*

FLAME RETENTION HEAD OIL BURNERS

Advances in oil burner technology are centered around flame retention head oil burners. They are designed to mix the air and oil to accomplish complete combustion of the fuel. These burners are identified by the retention head that replaces the traditional end cone. The retention heads are designed to create turbulence at the point of contact of the air and oil as they leave the burner blast tube. There are several variations that follow:

1. The internal cup and end cone was the first version (see Figure 3-18). The cup fills most of the interior of the blast tube that forces the air from the fan to the outer edge of the end cone as it leaves the burner. This results in an air delivery pattern that resembles a sunflower. The flame is held close to the edge of the burner by this type of air pattern. The shape of the flame can be adjusted by moving the cup back from the end cone. This allows more air to flow toward the center of the blast tube, reducing to some extent, the sunflower shape of the flame.
2. The fixed retention heads are mounted on the end of the blast tube (see Figure 3-19). Each head is designed for a small range of nozzle sizes. Because of this, a burner that can fire over a range of .50 to 2.50 gph can have five or six different heads available for installation. There is a matching internal pressure plate for each head size. The pressure plate is designed to force air to the outer edge of the blast tube. This directs the air to the retention head slots and vanes creating a small low pressure area directly in front of the nozzle. This reduced pressure area allows air to come back into the oil

Figure 3-18   **The Shell retention head is still found on many older oil burners.**

Figure 3-19   **Beckett AF Flame Retention Head oil burner.** *Courtesy: R. W. Beckett Corp.*

spray as it leaves the nozzle. The cup shape of the head and the air movement will hold the flame close to the head, provide excellent mixing of air and oil, and produce a compact, hot flame.

3. The adjustable flame retention head burners have an end cone and ring with vanes in them attached to the drawer assembly of the burner (see Figure 3-20). The vaned ring can be moved back against the end cone by moving the drawer assembly back. This is the full choke position used for the smallest nozzle firing rate for the partic-

Figure 3-20 **Wayne Blue Angel Flame Retention Head oil burner can operate under positive draft conditions.** *Courtesy: Wayne Home Equipment.*

ular burner. As larger nozzle sizes are required, the drawer assembly is moved forward, which creates a gap to form between the outer edge of the vaned ring and the end cone, allowing more air to enter the combustion process. This type of adjustable retention head will create the same low pressure area in front of the nozzle, leading to air movement back into the oil spray. The resulting flame is held close to the head giving us a hot, compact flame.

FORCED DRAFT OIL BURNERS

As boiler manufacturers search for higher efficiencies, the boilers have become more and more compact. This has led to restricted flue gas passage through the boiler that creates a problem for the burner. The combustion process releases a large volume of hot gas. If this gas remains in the fire box area of the boiler, the oil burner cannot supply the air necessary to continue the combustion process. The forced draft burners can push the flue gas out of the fire box area into the flue gas passages, and out of the boiler into either a chimney or vent to the outside of the building.

In order to accomplish this, these burners are all equipped with 3,450-rpm motors. The blower wheels, or fans, are as large as possible leaving little clearance to the interior of the burner chassis. There is an additional shroud made of sheet metal attached to the chassis that fits inside the fan (see Figure 3-21). This directs the air to the fan blades

Figure 3-21 **Fan augmentation shroud inside the Carlin EZ-1 burner chassis.** *Courtesy: Carlin Combustion Technology, Inc.*

and does not allow any of the air to recirculate back into the open end of the fan. The burner chassis has a small slot for the air leaving the fan to enter the burner blast tube (see Figure 3-22). This maintains the velocity achieved by the fan and sends the air, under pressure, through the flame retention head into the combustion chamber.

Please understand that while most of the forced draft burners now in use for residential oil burners are of the flame retention type, not all flame retention head burners are forced draft burners. Forced draft burners should only be used when positive draft conditions in the boiler cannot be overcome by the chimney. Draft conditions and their effects on burner operation are discussed in great detail in Chapters 4 and 20.

Carlin EZ-1

The Carlin EZ-1 oil burner is a prime example of a forced draft burner. This burner can be found on new equipment and can also be used as a replacement burner to upgrade older units (see Figure 3-23). The standard burner chassis has been modified with the addition of a fan augmentor. This is a 2″ wide strip of metal that fits inside the fan. It directs air into the fan blades as they rotate toward the blast tube. The opening into the blast tube has been reduced in size as well. This will help maintain the air velocity necessary to overcome positive pressure in the fire box of the boiler or furnace. The fan has large concave blades capable of moving twice the amount of air necessary for combustion of the fuel delivered through the rated nozzle sizes.

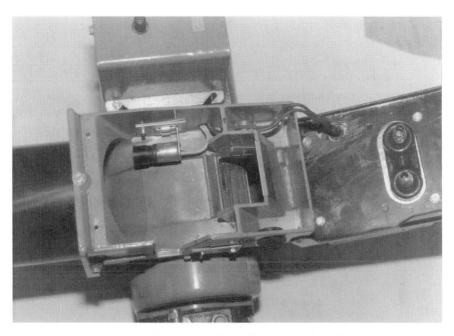

Figure 3-22 **Wayne Blue Angel chassis' interior has a small opening for air to leave the fan and enter the blast tube. This will provide the air velocity required to operate the burner under positive draft conditions.** *Courtesy: Wayne Home Equipment.*

Figure 3-23 **The Carlin EZ-1 forced draft oil burner.** *Courtesy: Carlin Combustion Technology, Inc.*

The outstanding feature of the EZ-1 is the unique combustion head (see Figure 3-24). The head is composed of two separate parts—the flat retention head mounted on the drawer assembly and the air distributor mounted on the end of the blast tube. The air distributor has a choke ring on the inside and a series of openings around the outer edge. The combination provides a precise air delivery pattern that holds the flame close to the head. The position of the retention head in relation to the choke ring is extremely important. As the firing rate is increased, the retention head must be moved forward to open the space between the head and the choke ring. This will allow additional air to move into the combustion process by passing around the retention head. To make this setting precise, a set of plates is sent with every EZ-1 burner (see Figure 3-25). The plates are installed on the pump side of the burner chassis. Each of the plates is stamped with one of the seven nozzle sizes that the EZ-1 can fire. The plate has an opening for the drawer assembly oil pipe where the tubing from the pump is connected. When the plate with the firing rate of the nozzle installed in the burner is used, you can be certain that the retention head is in the correct position inside the choke ring.

Beckett AF II

The Beckett AF II (see Figure 3-26) another forced draft burner is usually part of a boiler-burner package. It is particularly well suited for direct vent systems because the air intake shield can be removed and a 4″ air duct can be connected directly to the side of the chassis. The motor is a newly designed permanent split capacitor type equipped with ball bearings. The fan is mounted on the motor shaft close to the flat face of the motor (see Figure 3-27). The fuel pump is mounted on the rear end bell of the motor. The pump

Figure 3-24   **Carlin EZ-1 combustion head.** *Courtesy: Carlin Combustion Technology, Inc.*

Figure 3-25 **The Carlin EZ-1 is shipped with a set of plates to properly position the drawer assembly for the nozzle firing rate.** *Courtesy: Carlin Combustion Technology, Inc.*

Figure 3-26 **The Beckett AF II forced draft oil burner.** *Courtesy: R. W. Beckett Corp.*

Figure 3-27 **Beckett AF II motor and fan assembled.** *Courtesy: R. W. Beckett Corp.*

shaft fits into the rotor of the motor. A plastic bushing is used to absorb the starting shock (see Figures 3-28 and 3-29).

The chassis design (see Figure 3-30) maintains air velocity and sends the air spiraling down the blast tube to the retention head. The air adjustment for the burner is on the discharge side of the fan. This helps to maintain the air velocity as well. There are two different blast tubes available for the AF II oil burners. The first has a fixed retention head mounted on the end of the blast tube. The second has an adjustable flat retention head mounted on the drawer assembly with a choke ring set back from the end of the blast tube. To ensure proper placement of the retention head in the choke ring, a set of positioning pins (see Figure 3-31) is sent with every burner. A pin is installed in the drawer assembly as shown in Figure 3-32. When the drawer assembly is placed in the burner it is pushed forward until the positioning pin engages the rear of the choke ring. The smallest nozzle size needs the longest pin so that the retention head will remain close to the choke ring. As larger nozzle sizes are required, shorter pins are used. This allows the retention head to move forward and create a space between the retention head and the choke ring. More air can now flow into the combustion chamber, some of it through the vanes in the retention head and some that goes around the retention head. Detailed instructions regarding the proper pin to use for each possible nozzle size, as well as additional critical information about the installation of the burner, are always included in the original equipment instruction manual. It is extremely important for technicians to read all the instructions sent with the equipment before attempting to install or service the burner.

Figure 3-28 **Beckett AF II motor showing rear end bell pump mounting opening.** *Courtesy: R. W. Beckett Corp.*

Figure 3-29 **Fuel pump for Beckett AF II with plastic bushing that fits into the motor rotor.** *Courtesy: R. W. Beckett Corp.*

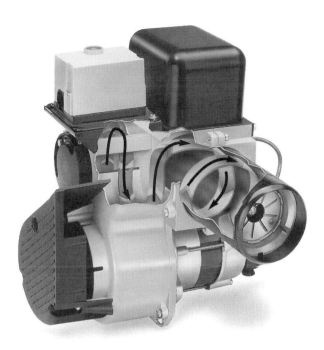

Figure 3-30 **Cutaway view of Beckett AF II chassis shows how air flows through the burner.** *Courtesy: R. W. Beckett Corp.*

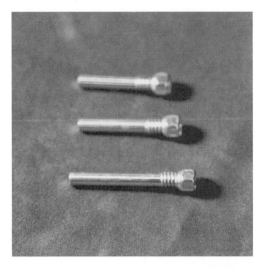

Figure 3-31 **Positioning pins for Beckett AF II drawer assembly.** *Courtesy: R. W. Beckett Corp.*

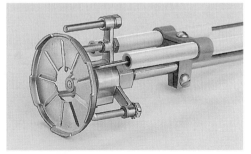

Figure 3-32 **Positioning pin installed in Beckett AF II burner drawer assembly.** *Courtesy: R. W. Beckett Corp.*

Riello F Series

The Riello F Series forced draft oil burners bring European technology to North America (see Figure 3-33). These burners can be used as replacements to upgrade existing boilers or as original equipment on new units. They are easily recognized by the acoustically insulated steel cover that encloses the working parts of the machine. Once the cover is removed, the parts become easily accessible (see Figure 3-34).

Some Features of the F Series Burners Are:

1. The ball bearing, permanent split capacitor motor generates the power for the primary control.
2. The primary control contains an electronic ignitor that delivers an 8,000-volt spark to the electrodes. The electrode setting (see Figure 3-35) is very important for the smooth light off of the burner flame.
3. The flat retention head is mounted on the burner drawer assembly. The space between the retention head and the choke ring is adjusted by using the guide on the burner chassis (see Figure 3-36). As the nozzle size is increased, the retention head is moved forward to open a space between the head and the choke ring. This allows additional air to enter the combustion process.

Figure 3-33 **Riello F-3 oil burner.** *Courtesy: Riello Corporation of America.*

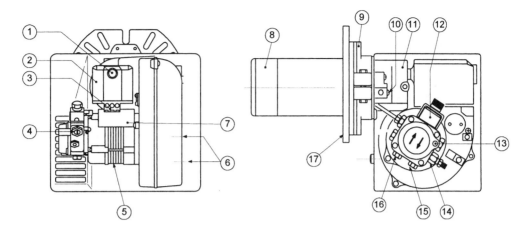

F3 AND F5 BURNER COMPONENTS

1. Lockout indicator lamp and reset button
2. Primary control
3. Primary control sub-base
4. Pump pressure regulator
5. Motor
6. Air adjustment fixing screws
7. Capacitor
8. Combustion head
9. Semi flange
10. Turbulator adjustment screw

11. Air tube cover
12. Coil
13. Vacuum gauge port
14. Pressure gauge and bleeder port
15. Return fuel line port
16. Supply fuel line port
17. Mounting flange with gasket

Figure 3-34 **Major component parts of the Riello F Series oil burner.** *Courtesy: Riello Corporation of America.*

IMPORTANT: THESE DIMENSIONS MUST BE OBSERVED AND VERIFIED.

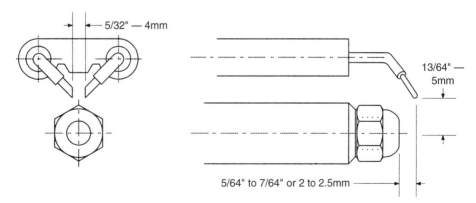

Figure 3-35 **Electrode setting for Riello F3 and F5 oil burners.** *Courtesy: Riello Corporation of America.*

A) Loosen **nut** (1) then turn **screw** (2) until the **index marker** (3) is aligned with the correct index number as per the guide on the burner chassis.
B) Retighten the **retaining nut** (1).

MODEL F5 NOTE: Zero and 4 are scale indicators only. From left to right, the first line is 4 and the last line 0.
MODEL F3: Same as above, except scale indicators are 0 and 3.

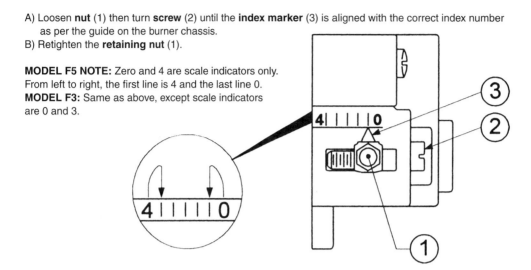

Figure 3-36 **Directions for adjusting the position of the retention head in the choke ring by using the guide on the burner chassis.** *Courtesy: Riello Corporation of America.*

4. There is an automatic air shutter controlled by a hydraulic piston. When the burner is off, the air shutter is closed. When the burner starts and the fuel pump pressure reaches 80 psi, the piston moves the shutter to a preset open position.
5. The fuel pump has a built-in solenoid valve to control the flow of oil to the nozzle and air piston. When the solenoid is in the open position, oil bypasses the pressure regulator and recirculates in the strainer of the pump. When the solenoid closes, the oil is sent through the pressure regulator, opening the air shuttle at 80 psi. The pressure continues to rise to 140 psi, spraying oil through the nozzle into the combustion chamber where it mixes with air and ignites.
6. The pump has metric sized British Parallel Threads (BPT) in all of its ports (see Figure 3-37). To connect the oil lines or install standard gauges, adaptors have to be used to convert the BPT to National Pipe Threads (NPT). A set of adaptors is sent with every new oil burner.

The Sequence of Operation for the F Series Burners Is as Follows:

1. On a call for heat, the burner motor starts, creating an air flow through the unit.
2. The primary control is powered and the electronic ignitor establishes a spark at the electrode tips.
3. The solenoid valve closes, the air shutter opens, and oil is sprayed through the nozzle into the combustion chamber.
4. The air–oil mixture is ignited and the cad cell flame detector senses the flame.

OIL LINE CONNECTIONS

This burner is shipped with the oil pump set to operate on a **single** line system. To operate on a **two-line** system the bypass plug (4) must be installed.

WARNING: Do not operate a **single** line system with the bypass plug installed. Operating a **single** line system with the bypass plug installed will result in damage to the pump shaft seal.

NOTE: Pump pressure must be set at time of burner start-up. A pressure gauge is attached to the **pressure port** (8) for pressure readings. Two **pipe connectors** (6) are supplied with the burner for connection to either a single or two-line system. Also supplied are two **adaptors** (3), two female 1/4" NPT, to adapt oil lines to burner pipe connectors. All pump port threads are **British Parallel Thread** design. Direct connection of NPT threads to the pump **will damage** the pump body. (5 and 7) Riello manometers and vacuum gauges **do not** require any adapters, and can be safely connected to the pump ports. An NPT (metric) adapter **must** be used when connecting other gauge models.

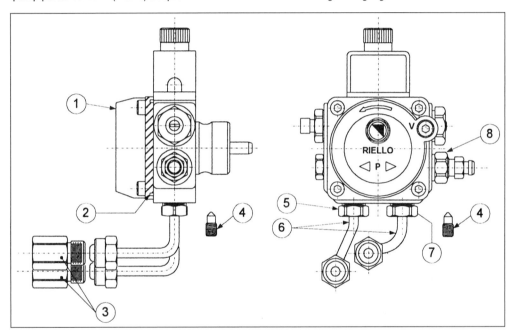

Figure 3-37 **Riello fuel pump connections. Adapters must be used to convert BPT to NPT.**
Courtesy: Riello Corporation of America.

5. The electronic ignitor shuts off and the burner runs until the call for heat is satisfied. Then the oil solenoid opens, the flow of oil to the nozzle stops, and the flame goes out. The air shutter closes, the motor stops, and the burner sits quietly waiting for the next call for heat.

The oil burners described in this chapter are only part of the equipment now in operation in the millions of homes heated by oil. The information is introductory in nature. Specific, detailed instructions on a particular burner can be obtained from the manufacturer. Many oil burner companies conduct training seminars for technicians working in the industry. It is extremely important to attend these meetings as well as trade shows to keep up with the changes in technology and equipment.

We can list the parts belonging to each particular system built into high pressure gun type oil burners:

The Ignition System

Ignition transformer
Electrodes (2)
Bus bars or high tension wires (2)
Electrode bracket

The Fuel Delivery System

Motor
Fuel pump coupling
Fuel unit (pump)
Nozzle
Nozzle adaptor
Copper tubing
Delayed opening oil valve

The Air Delivery System

Oil burner chassis
Motor
Blower wheel or fan
Blast tube
Air intake shutter
Stabilizer
End cone
Flame retention head
Forced draft internal shroud

To Briefly Summarize

1. An oil burner is a machine that prepares liquid fuel for combustion, starts the combustion process, and projects the resulting flame into a boiler or furnace in which the heat released can be utilized.
2. Every oil burner must have an air delivery system, a fuel delivery system, and an ignition system.
3. Oil burners are classified by the method used to prepare the fuel for combustion.
4. The high pressure gun type oil burners dominate the residential oil burner market.
5. Advances in oil burner technology in recent years have centered around the flame retention head oil burners.

Please Answer These Questions

1. How are oil burners classified?
2. Identify the three separate systems that must be built into every oil burner.
3. Which type of oil burner requires preheating of the oil and why is it necessary?
4. Explain primary and secondary air.
5. Why did the high pressure gun type oil burner become so popular for residential use?
6. Identify the three types of flame retention head oil burners.
7. How has the flame retention head improved the efficiency of residential oil burners?
8. What advantage does the adjustable retention head have over the fixed heads?
9. What is the difference between a forced draft burner and a flame retention head burner?
10. When should forced draft burners be used?

CHAPTER 4

THE AIR DELIVERY SYSTEM

The air delivery system of the high pressure gun type oil burner has three functions:

1. To supply the air required for the complete combustion of the fuel.
2. To mix the air with the oil to provide a uniform, combustible mixture that will burn completely.
3. To provide a method for the removal of the flue gas, the hot gas formed by the combustion process, from the boiler.

FRESH AIR LOUVRES

The oxygen for combustion comes from the air that we breathe. Since only 21% of air is oxygen (the remaining 79% is nitrogen), we need a large volume of air to complete the combustion process. Actually, 1,540 cubic feet of air is required to completely burn one gallon of fuel oil.

Many boilers and furnaces are installed in open basements where there is an adequate supply of infiltration air available in the building through openings around windows and doors for combustion. The addition of storm doors and windows and weather stripping to a building severely restricts the infiltration air supply. This makes the installation of a fresh air louvre near the heating equipment necessary (see Figure 4-1). Local building codes may determine the size of fresh air openings for oil burners. When there is no applicable local code requirement, we suggest using the information regarding fresh air intake found in the National Fire Prevention Association publication NFPA 31 as outlined below.

The opening size for a louvre that allows air into a building for a boiler operating in an unconfined space can be computed by using the following formula (see Figure 4-2). The louvre slots will block approximately half the air so the actual size of the louvre frame will be twice the square inches calculated by using the formula.

Total Free Area (TFA) = 1 square inch per 5,000 Btus/hour of total input rating of all units

Example: There is a heating boiler, input rating 130,000 Btus, and a hot water heater, input rating 70,000 Btus, operating in a basement. Compute the louvre size for this building.

TFA = 130,000 + 70,000 / 5,000

TFA = 200,000 / 5,000

TFA = 40 square inches

Louvre Size = 40 × 2

Louvre Size = 80 square inches

When the boiler is in an enclosed space in a building with adequate infiltration air available for combustion, there must be two openings to the space, one near the ceiling and the other near the floor (see Figure 4-3). The formula for computing the size of each of these openings is as follows:

TFA = 1 square inch per 1,000 Btus/hour of input rating

Example: A boiler room contains a heating boiler, input rating 90,000 Btus, and a hot water heater, input rating 50,000 Btus. Compute the size of the openings required for this boiler room.

TFA = (90,000 + 50,000) / 1,000

TFA = 140,000 / 1,000

TFA = 140 square inches

When the boiler is enclosed in a space in a building that does *not* have adequate infiltration air available for combustion, there must be two openings to the outside air, one

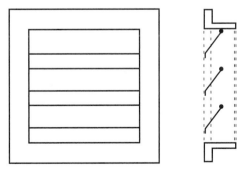

Figure 4-1 **Basic fresh air intake louvre is mounted in basement outside wall near the boiler to supply continuous fresh air to the oil burner.**

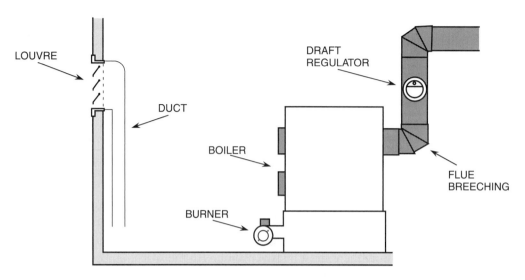

Figure 4-2 **Louvre with duct to direct fresh air to the basement floor to eliminate the possibility of freezing water piping. Duct must not reduce the total free area required.**

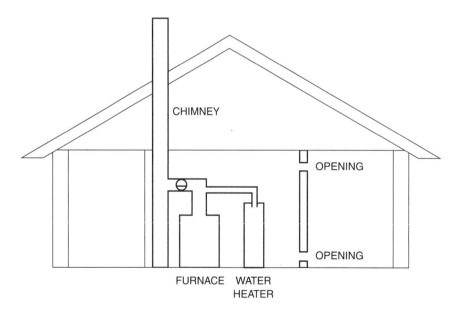

Figure 4-3 **Air openings for heating equipment in a confined space. The air for combustion is from inside of the building.**

near the ceiling and the other near the floor (see Figure 4-4). The size of each of these openings can be computed by using the following formula.

TFA = 1 square inch per 2,000 Btus/hour of input rating

Example: A boiler room contains a furnace with an input rating of 110,000 Btus. Compute the size of the openings required for this boiler room.

TFA = 110,000 / 2,000

TFA = 55 square inches

Remember, the requirement is for total free area of air intake. Provisions must be made so that louvres, vent hoods, and ductwork do not reduce the required size.

Figure 4-5 illustrates a system in which the oil burner has a permanent duct connection to the outside air. This duct must contain a vacuum breaker (a barometric draft regulator can be used for this purpose) to allow air from inside the building to be drawn into the burner if the outside duct is blocked. The unit in this illustration is also equipped with a direct vent instead of having a connection to a chimney. We will discuss chimneys and direct venting later on in this chapter.

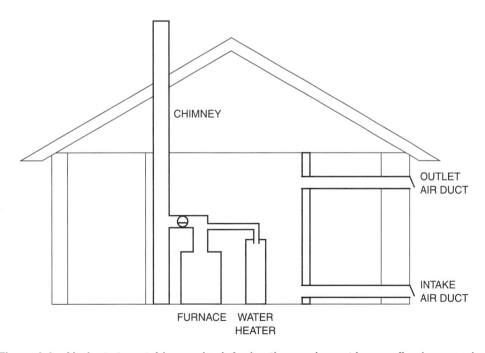

Figure 4-4 Air ducts to outside supply air for heating equipment in a confined space when an adequate supply of air is not available from inside the building.

Figure 4-5 **Direct vent system with outside air duct connected to oil burner air intake opening.** *Courtesy: Burnham Corporation.*

In many older buildings, the heating unit was installed in a large basement. The oil burner in this type of installation had an adequate supply of infiltration air. Once the unit is enclosed in a boiler room, it becomes necessary to ensure an adequate air supply. When an outside louvre is not practical, there must be openings near the bottom and top of the enclosure as in Figure 4-3. Each of these openings must have a free area of 140 square inches per GPH used by the burner. As an example, if the firing rate of the burner was 1.50 GPH, each opening should be 1.50 × 140 = 210 sq. in. If there is an oil-fired hot water heater in the same enclosure, that burner's firing rate must be included in the calculations. As an example:

$$\text{heating system burner} = 1.50 \text{ GPH}$$

$$\text{hot water heater burner} = .75 \text{ GPH}$$

$$1.50 + .75 = 2.25 \text{ GPH}$$

$$2.25 \times 140 = 315 \text{ sq. in.}$$

An increasingly popular method of supplying combustion air is to attach a duct from the air intake openings on the burner chassis directly through an outside wall as in

Figure 4-4. The end of the duct must be located high enough on the wall so that it does not become blocked by snow, leaves, or other debris. A vacuum breaker, usually a draft regulator, is installed in the duct so that if the outside end of the duct is blocked, the draft regulator will open and allow interior building air to enter the burner. There are units made as in Figure 4-5 that allow for sidewall venting of the flue gas as well as fresh air intake for the oil burner.

   A constant source of fresh air is crucial to the operation of any oil burner. When the combustion process does not get sufficient oxygen, the result is smoke and soot. Homeowners do not appreciate smoke and soot and will go to great lengths to rid themselves of the problem. Their solution could be switching to another supplier with a service department that understands the combustion process and can get the burner to operate properly.

AIR DELIVERY SYSTEM PARTS

Oil burners are rated according to the amount of fuel, in gallons per hour, that they can burn. Burners are designed to operate within a range of sizes so that one model can be used for a variety of installations. The determining factor in the rating of burners is the capacity of the air delivery system. The parts and their function are listed below:

1. The main burner housing, or chassis, is in the shape of a scroll. This shape allows for the smooth movement of air without any turbulence inside the housing. On one side of the housing is a large, carefully machined opening where the motor is mounted. On the side of the housing opposite the motor is a smaller opening where the fuel pump is mounted. The air intake slots and adjusting band or shutter are on the side of the burner near the fuel pump (see Figure 4-6).

2. The fan is mounted on the motor shaft (see Figure 4-7). The fan is made of a solid plate that has a mounting hub in its center. Attached to this plate is a series of concave blades that push the air as the motor rotates. The blades are held together by a ring on the end opposite the hub plate. There should be as little clearance as possible between the fan and the chassis. This will ensure the necessary velocity for the delivery of air to the blast tube.

3. The blast tube extends from the discharge end of the chassis scroll. The burner gets its name, *gun type*, from the way the blast tube and scroll assembly resemble a gun (see Figure 4-8). The drawer assembly, containing the oil burner nozzle and electrodes, fits inside the tube. The end of the tube is set 1/2″ back from the interior wall of the combustion chamber (see Figure 4-9).

4. At the end of the blast tube is the end cone (see Figure 4-10). The end cone reduces the size of the blast tube opening and may have vanes inside to cause the air to change direction as it leaves the tube. Restricting the air at the end of the tube will cause it to accelerate. This is commonly known as a Venturi effect. Deflecting the air to make it change direction at this same point will create the turbulence required to mix the air and oil as they leave the burner.

Figure 4-6 **Aero F-AFC oil burner air intake adjustment band.** *Courtesy: Aero Environmental Ltd.*

Figure 4-7 **Oil burner fan—sometimes called the blower wheel.** *Courtesy: Carlin Combustion Technology, Inc.*

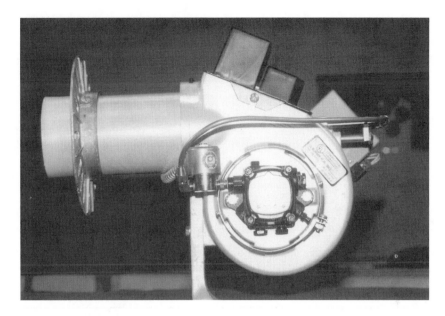

Figure 4-8 **Carlin 201 CRD oil burner.** *Courtesy: Carlin Combustion Technology, Inc.*

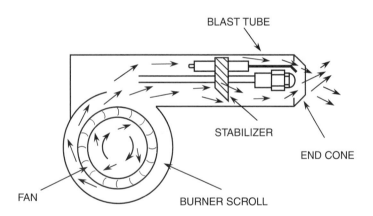

Figure 4-9 **Standard high pressure gun type oil burner air delivery system with stabilizer and end cone.**

5. The stabilizer is part of the drawer assembly (see Figure 4-11). It acts as the mounting bracket for the electrodes. It keeps the drawer assembly centered in the blast tube. The shape of the stabilizer forces the air moving through the blast tube to change direction and accelerate as it approaches the end cone. It is essential that the stabilizer and the vanes on the end cone turn the air in the same direction. If they do not, the acceleration is lost and the result is improper mixing of the air and oil.

Figure 4-10 **Standard oil burner end cone.**

Figure 4-11 **Standard oil burner stabilizer is also used as the electrode bracket.**

SHELL HEAD BURNERS

The parts described above are found in standard gun type oil burners. There are some variations that are intended to improve the mixing of the air and oil. The first of these is called the *Shell head* (see Figure 4-12). In this system, the stabilizer is replaced by a cup that fills most of the interior of the blast tube. There is an adjustment that allows the cup to be moved closer or farther from the end cone of the burner. This movement will either open or close the opening at the end of the blast tube depending on the amount of air required to burn the oil coming out of the nozzle. Because the cup fills the center of the blast tube, the air coming down the tube is forced to the outer surface of the tube. This

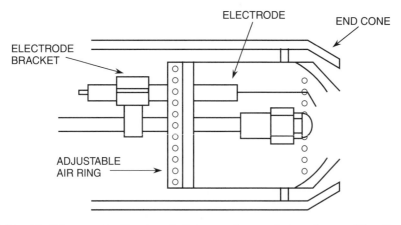

Figure 4-12 **Shell flame retention head oil burner has an internal cup with adjustable air slots that fits into the burner blast tube.**

leaves the center of the tube with little or no air moving through it. The result is an air delivery pattern that resembles a sunflower with most of the air on the outer edges and little in the middle. The air pattern can be adjusted by opening the air slots at the rear of the cup and moving the edge of the cup back from the burner end cone.

FLAME RETENTION HEAD OIL BURNERS

The most recent development in improving the air–oil mixture is the flame retention head (see Figures 4-13 and 4-14). In this setup the stabilizer is replaced by a pressure plate mounted on the electrode bracket. This is a flat disc that blocks part of the blast tube, causing the air to accelerate as it passes. The burner end cone has been replaced by the retention head. There are two types of retention heads: the fixed and the adjustable. The fixed heads attach to the end of the blast tube. They are made of heat resistant steel in the shape of a flat cup. There is a center opening for the oil spray and a series of slots around the interior of the head. The size of the slots varies with the amount of oil to be fired by the nozzle. One size head can be used for a specific range of nozzle sizes. It is necessary to install the proper size head and matching pressure plate when the burner is installed (see Figure 4-15).

The adjustable version has a retention head that fits on the end of the drawer assembly. It has a center opening for the oil spray and a series of vanes around that opening. The outer edge of the vane assembly fits against an interior ring in the blast tube that seals the edge of the head. The adjustment, according to the nozzle size, is accomplished by moving the head away from the interior ring. This opens the air passage around the head and increases the amount of air entering the air–oil mixture. The fully closed position, known as full choke, would be for the smallest nozzle size in the burner range (see Figures 4-16 and 4-17).

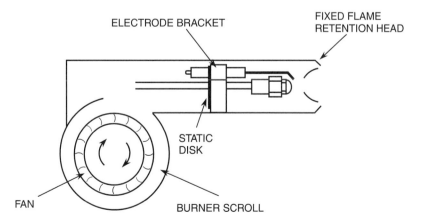

Figure 4-13 A flame retention head oil burner has a static disc inside the blast tube. The disc and the retention heads are matched for the proper air delivery to the combustion process.

Figure 4-14   This Wayne flame retention head oil burner has the retention head attached to the drawer assembly so that it can be easily cleaned. *Courtesy: Wayne Home Equipment.*

Unitized Flame Retention Heads

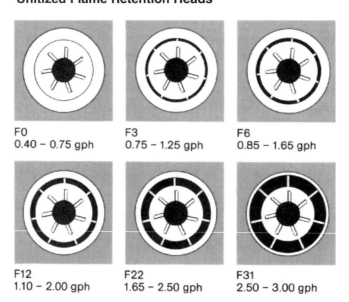

F0
0.40 – 0.75 gph

F3
0.75 – 1.25 gph

F6
0.85 – 1.65 gph

F12
1.10 – 2.00 gph

F22
1.65 – 2.50 gph

F31
2.50 – 3.00 gph

Figure 4-15   Flame retention heads for the Beckett AF oil burners are rated in gallons per hour of firing rate. Notice that as the firing rate increases, the air opening slots get larger. *Courtesy: R. W. Beckett Corp.*

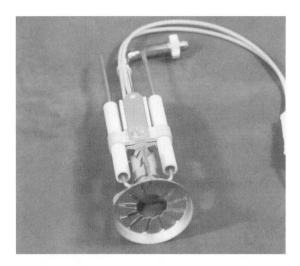

Figure 4-16 **The Carlin EZ-1 burner has the retention head mounted on the drawer assembly.** *Courtesy: Carlin Combustion Technology, Inc.*

Figure 4-17 **The Carlin EZ-1 burner flame retention head fits into the unique air distributor end cone.** *Courtesy: Carlin Combustion Technology, Inc.*

The shape of the retention head leaves a space with little air in the center of the air delivery pattern. This allows air from the outer edge of the pattern to turn back toward the burner where it can create additional turbulence and mix with the oil coming out of the burner nozzle. This action holds the flame at the face of the retention head that results in a compact, intensely hot flame.

Most of the flame retention head burners are equipped with motors that run at 3,450 rpm. Increasing the speed allows the use of smaller fans and burner housings. The increased velocity of the air moving through the retention head improves the air–oil mixture and leads to high efficiency operation. There are some variations, such as the Aero FAFC burner that uses a large fan and 1,725-rpm motor for flame retention and quiet operation. Another variation is the Adams INTERburner which features a reversed air system (see Figure 4-18). The INTERburner also features a ceramic end cone that gets hot to help vaporize the oil as it leaves the nozzle (see Figure 4-19).

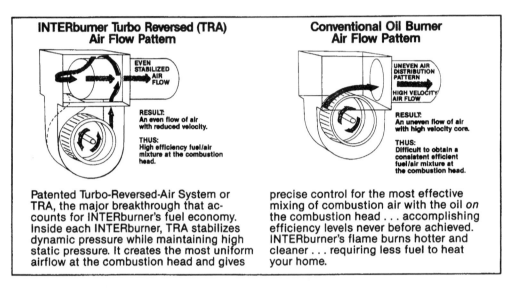

INTERburner Turbo Reversed (TRA) Air Flow Pattern

EVEN STABILIZED AIR FLOW

RESULT: An even flow of air with reduced velocity.

THUS: High efficiency fuel/air mixture at the combustion head.

Patented Turbo-Reversed-Air System or TRA, the major breakthrough that accounts for INTERburner's fuel economy. Inside each INTERburner, TRA stabilizes dynamic pressure while maintaining high static pressure. It creates the most uniform airflow at the combustion head and gives

Conventional Oil Burner Air Flow Pattern

UNEVEN AIR DISTRIBUTION PATTERN

HIGH VELOCITY AIR FLOW

RESULT: An uneven flow of air with high velocity core.

THUS: Difficult to obtain a consistent efficient fuel/air mixture at the combustion head.

precise control for the most effective mixing of combustion air with the oil *on* the combustion head . . . accomplishing efficiency levels never before achieved. INTERburner's flame burns hotter and cleaner . . . requiring less fuel to heat your home.

Figure 4-18 **Unique air flow through the Adams INTERburner.** *Courtesy: Adams Manufacturing Co.*

Figure 4-19 **The Adams INTERburner features a ceramic end cone to help vaporize the oil as it leaves the nozzle.** *Courtesy: Adams Manufacturing Co.*

FORCED DRAFT OIL BURNERS

The quest for higher efficiency has led the boiler and furnace manufacturers to compact units with restricted flue gas passages. These smaller units require a burner that can overcome positive over-the-fire-draft conditions. A standard oil burner, even a flame retention head burner, cannot develop the static air pressure required to force combustion air into a positive draft situation. This is because if the end of the blast tube is blocked the air will "spill over" the fan as it turns and remain in the burner chassis. The result will be a smoky, sooty flame and eventually a safety lockout. Forced draft burners are designed to overcome positive draft conditions (see Figure 4-20).

The Carlin EZ-1 is a forced draft burner. A standard burner chassis has been modified by adding a fan augmentor and significantly reducing the size of the opening between the burner scroll and the blast tube (see Figure 4-21 and 4-22). The fan augmentor is a flat metal plate that fits inside the fan. When the burner is running, the augmentor directs air into the fan blades as they turn up toward the blast tube. This simple device prevents spill over and increases the static pressure developed by the fan approximately 40%. Restricting the size of the air opening into the blast tube maintains the static pressure and velocity of the air necessary to overcome the positive draft in the boiler fire box. An important point to remember is that all forced draft burners are of the flame retention type but not all flame retention burners are of the forced draft type. Check the manufacturer's specification sheet or instructions sent with the burner before you install or service a burner if you are not certain which type you are dealing with.

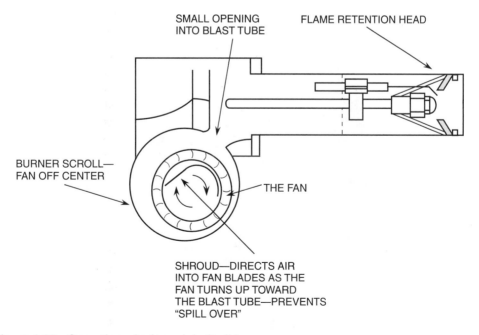

Figure 4-20 **Operation of a forced draft oil burner.**

Figure 4-21   **The fan augmentor in the Carlin EZ-1 chassis fits inside the fan. It directs air to the spinning fan blades and prevents air "spill over" inside the scroll.** *Courtesy: Carlin Combustion Technology, Inc.*

Figure 4-22   **Reducing the size of the air opening to the blast tube maintains static air pressure produced by the fan.** *Courtesy: Carlin Combustion Technology, Inc.*

SETTING THE RETENTION HEAD

Flame retention head burners, as well as forced draft burners, have a critical dimension: the distance between the face of the nozzle and the outer edge of the retention head itself (see Figure 4-23). Included in the installation instructions shipped with every new burner regardless of manufacturer is a detailed method of achieving the proper setting. Burners with fixed retention heads must be set before the burner is installed (see Figure 4-24). Once the setting is completed, usually by using a gauge sent with the burner, the burner chassis should be marked to indicate the exact location of the drawer assembly. This will ensure that the proper spacing is set when the burner is taken apart for servicing and then reassembled. The retention heads that are part of the burner drawer assembly

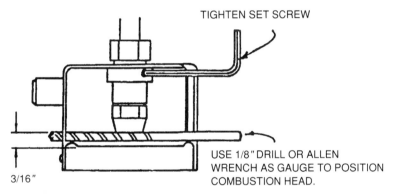

Figure 4-23 **Positioning the retention head of an Adams INTERburner.** *Courtesy: Adams Manufacturing Co.*

Figure 4-24 **Using the Beckett T-gauge to set the position of the nozzle in relation to the retention head. This must be done before the burner is installed.** *Courtesy: R. W. Beckett Corp.*

must be set as well. The manufacturer's instructions must be followed when the burner is first installed. To ensure the correct spacing, some of the manufacturers of removable retention heads have attached them to the nozzle adaptor so that the exact spacing is achieved and cannot be changed. Burner manufacturers also recognize the importance of the exact placement of the drawer assembly retention head in relationship to the interior choke ring at the end of the blast tube. The Carlin EZ-1 burner is shipped with a set of 7 plates that position the head according to the firing rate of the nozzle installed in the burner. Each plate has a nozzle size stamped on it. When the plate matching the exact nozzle size is installed, the retention head will be in the exact setting to allow the essential air flow for the amount of fuel entering the combustion chamber. An important point to remember when working on burners of this type is that as the nozzle size increases, the drawer assembly must be moved forward to open the space between the retention head and the choke ring in the end cone or blast tube. Service technicians have to check the setting carefully when the burner is serviced so that it can be reassembled properly. The retention head is a vast improvement over the older end cone type burners. It is sad to see the results of improper settings that negate the improvement. You have a responsibility to learn how to set every type of burner you come into contact with. Do not be ashamed to ask questions, either at your local supply house or by writing directly to the manufacturer. They will be more than happy to help you with any problems you may be having.

AIR DELIVERY PATTERNS

The three types of burners described above have distinct air delivery patterns. Trained service technicians must be able to identify each pattern to ensure proper burner operation.

The Patterns Are:

1. Stabilizer and end cone = solid air delivery pattern (see Figure 4-25)
2. Shell head and end cone = hollow, sunflower air delivery pattern (see Figure 4-26)
3. Flame retention head = semisolid air delivery pattern (see Figure 4-27)

Air Delivery System Operation

To Recap the Operation of the Air Delivery System:

1. Fresh air enters the boiler room through the louvre.
2. When the burner starts, the fan pulls air through the air intake slots and pushes it through the blast tube.
3. The amount of air entering the burner is controlled by the air adjustment bands or shutter at the air intake slots.
4. The air moves to the end of the blast tube where the air handling parts cause acceleration and turbulence.
5. The air mixes with oil coming out of the nozzle. The mixture is ignited by an electric spark and the flame is established.

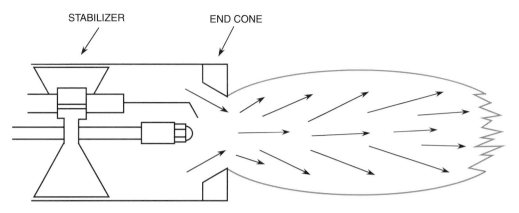

Figure 4-25 **Solid air delivery pattern delivered by burner with standard end cone and stabilizer.**

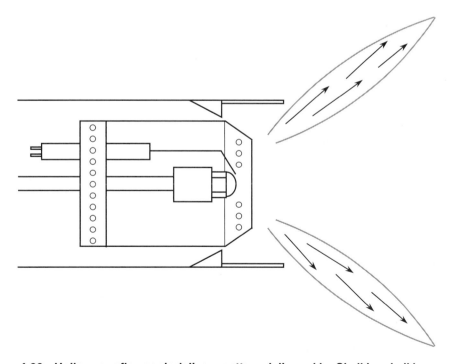

Figure 4-26 **Hollow, sunflower air delivery pattern delivered by Shell head oil burner.**

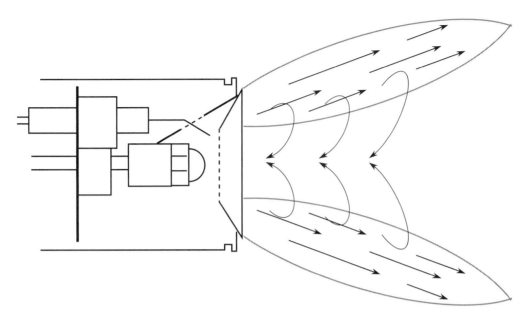

Figure 4-27   **Semi-solid air delivery pattern delivered by flame retention head oil burners.**

FLUE GAS REMOVAL

As soon as the flame starts, a large volume of hot gas (flue gas) is released. It is absolutely necessary for the flue gas to move through the boiler into the chimney and out into the atmosphere. If the flue gas does not continue to move, the oil burner will not be able to force air into the combustion process. The fan will rotate, but because there is no place for the air to go, the air will "spill over" the fan and stay in the burner chassis. This leads to incomplete combustion, smoke, soot, and eventual burner lockout on safety. Boilers and furnaces are designed to extract the heat from the flue gas. The efficiency of the boiler or furnace depends on how much heat is extracted. This heat exchange takes place in the flue gas passages. Soot build up inside the flue gas passages will insulate them and reduce the efficiency of the heat exchange.

The design of the flue gas passage varies with the unit manufacturer. The basic design has a fire box where the fuel burns at the bottom of the unit. The flue gas passages begin above the fire box and generally force the gas to travel back and forth before leaving through the flue gas outlet. The longer the flue gas remains in the unit, the more heat can be extracted and the higher the efficiency. New high-efficiency units equipped with forced draft burners can handle this with no problem. The flue gas is exhausted from the boiler or furnace into a vent that carries the gas to the outside of the building. The vast majority of oil-fired heating equipment now in operation needs help to get rid of the flue gas. That help is supplied by the chimney and it is called *draft*.

CHIMNEYS AND DRAFT

The construction and maintenance of chimneys is an ancient art. Since the beginning of recorded history men have known that if there is a controlled fire inside a building there must be a chimney to get the smoke out. Long before central heating systems were installed, fireplaces and wood burning stoves had chimneys. It is important for service technicians to understand the operation of the chimney (see Figure 4-28).

Draft can best be defined as a negative pressure or suction inside the boiler. You cannot feel draft in a boiler. You must use an instrument called a draft gauge to find out if there is indeed any pressure, negative or positive, inside the boiler. The draft gauge is based on a U tube that has water in it. When the base of the U is level, the water is at equal height in both legs of the U. Pressure exerted on the water in one of the legs would raise the level of the water in the other leg. Suction applied to the water in one of the legs would cause the water to drop in the other leg. If we connected one of the U's legs to a

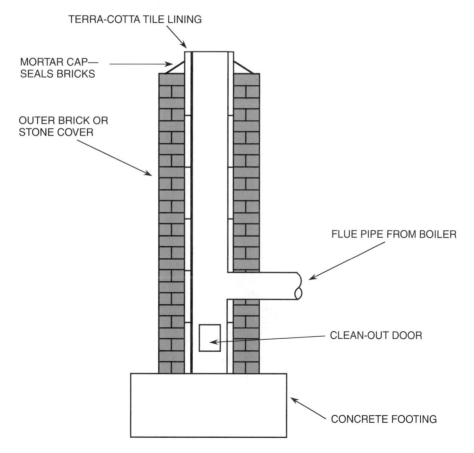

TERRA-COTTA TILE LINING

MORTAR CAP—
SEALS BRICKS

OUTER BRICK OR
STONE COVER

FLUE PIPE FROM BOILER

CLEAN-OUT DOOR

CONCRETE FOOTING

Figure 4-28 **Tile-lined chimney construction design.**

piece of rubber tubing and inserted the tubing into the flue gas passages of the boiler, we could see if the water in the other leg went up or down and exactly how much it moved. The unit of measurement for draft is inches of water column. The desired draft, measured over the fire with the burner in operation, is negative 2/100ths of an inch of water column. This is usually written as follows: –.02″ WC.

There are two distinct types of drafts: thermal and current. Thermal draft is created by the basic principle that heating a gas will make it rise. The flue gas entering the chimney is certainly hotter than the air around the outside of the chimney. This difference in temperature will start the flue gas up the chimney creating the thermal draft. Current draft is the result of air movement across the top of the open chimney. Obviously this will vary with the weather conditions. A clear, dry, windy day will produce a strong current draft while a damp, still day will not produce any draft at all (see Figure 4-29).

All chimneys can produce both types of drafts. In order to regulate the over-the-fire draft, a draft regulator is installed in the flue pipe that connects the flue gas outlet of the boiler to the chimney (see Figure 4-30). The draft regulator has an adjustable damper that opens to allow air from the boiler room to enter the chimney. The entry of this additional

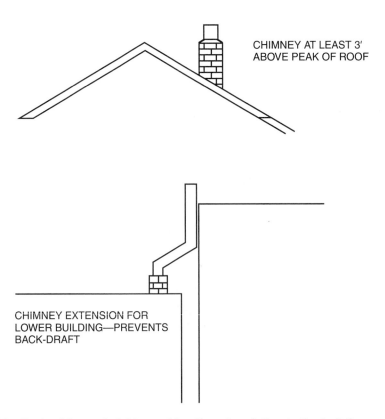

CHIMNEY AT LEAST 3′
ABOVE PEAK OF ROOF

CHIMNEY EXTENSION FOR
LOWER BUILDING—PREVENTS
BACK-DRAFT

Figure 4-29 **Basic chimney height considerations in relation to the buildings.**

Figure 4-30   **Field draft regulator.**   *Courtesy: Field Controls.*

cold gas will slow down the draft inside the boiler over the fire. The draft regulator
should be adjusted on a cool, windy day. The draft gauge must be used to ensure that
when the current draft is high, the draft over the fire will be the desired –.02″ WC. An
important point to remember is that the draft regulator can only *reduce* draft; it cannot
increase draft.

Chimneys should be constructed with a terra-cotta tile lining surrounded by bricks.
The lining may be round, (which works best), or square. Rectangular-shaped tile linings
do not work well and can cause serious problems (see Figure 4-31). The minimum dimensions for oil burner chimneys should be 7″ in diameter, or 8 × 8″ square, and 15′
high (see Table 4-1). Boiler and burner manufacturers' specifications must be followed
when installing equipment that requires a chimney.

Some Basic Recommendations for Chimney Construction Are:

1. Pay particular attention to the seams and joints. Cold air leaking into the chimney
   will reduce the draft.
2. There should be nothing on top of the chimney. Just leave the top open. Rain falling
   into the chimney will not hurt it under normal weather conditions.
3. The top of the chimney should extend at least 3′ above the top of the building.
4. There should be a clean-out door at the base of the chimney. This has to close tightly
   so that no air can leak into the chimney.
5. The chimney should run straight up with no offsets where dirt can collect.
6. Chimneys built inside buildings must be well insulated and sealed to prevent fires
   and smoke damage.

7. Building structural members must not extend into the chimney.
8. The flue pipe connection should not extend into the chimney, which would block it, and the pipe must be sealed with furnace cement to eliminate air leaks into the chimney. The flue pipe from the boiler outlet to the chimney must pitch or slant up toward the chimney.

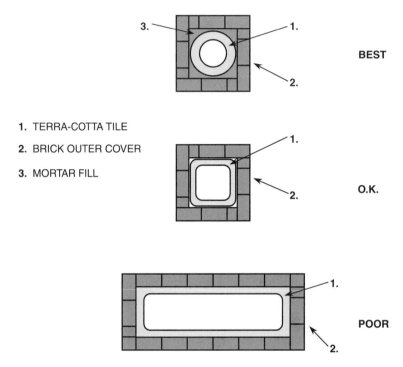

1. TERRA-COTTA TILE

2. BRICK OUTER COVER

3. MORTAR FILL

Figure 4-31 **Basic chimney shapes. Round is the best; square is acceptable, but rectangular is not good.**

GROSS BTU INPUT	RECTANGULAR TILE	ROUND TILE	MINIMUM HEIGHT
144,000	8 1/2″ × 8 1/2″	8″	20 ft
235,000	8 1/2″ × 13″	10″	30 ft
374,000	13″ × 13″	12″	35 ft
516,000	13″ × 18″	14″	40 ft
612,000	—	15″	45 ft
768,000	18″ × 18″	—	50 ft
960,000	20″ × 20″	18″	55 ft

Table 4-1 **Chimney Dimensions for Gross Btu Input of the Oil Burner.** *Courtesy: R. W. Beckett Corp.*

CHIMNEY LINERS

Flexible metal chimney liners are available to repair leaking masonry chimneys. They are also of use when an old oil burner is replaced with a new, high efficiency unit that has a forced draft burner. The old masonry chimneys are too large for these burners. They do not warm up enough to be effective, and create problems with the draft as well as allowing the flue gas to condense, which will damage the flue pipe and boiler breeching.

The flexible liner kit (see Figure 4-32) includes the flexible liner, flashing for the top of the masonry chimney, a support collar, a chimney rain cap, and a connector to attach to the flue pipe from the boiler or furnace. It is possible to bend the liner at the base of the chimney (see Figure 4-33) to make the connection to the boiler flue pipe. An alternative method is to use a tee in the chimney base (see Figure 4-34) to complete the connection.

Figure 4-32 **Chimney liner kit includes flexible pipe that is inserted into the chimney, flashing, a rain cap, and connectors.** *Courtesy: Z-FLEX.*

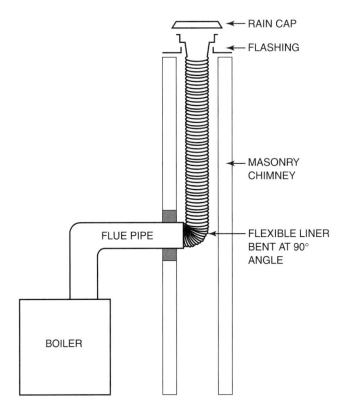

Figure 4-33 **The flexible liner can be bent at a 90° angle in the base of the chimney for the connection to the flue pipe.**

The liners are built in three sizes, 4, 5, and 6″. The required size depends on the height of the chimney, the horizontal length of flue pipe between the boiler and the chimney, and the input of the oil burner. The manufacturers of these kits have literature available to assist in selecting the correct liner for a particular job. The Liner Finder™ by Z-Flex (see Figure 4-35) is a slide rule type device that can help the installer determine the correct size liner for a particular chimney.

Metal chimneys have been developed to replace the conventional tile-lined version. These are much less expensive and easier to install. They are made of an inner tube surrounded by insulation and then covered with a weather-resistant steel jacket. The manufacturer's installation instructions regarding supports and flashing must be followed carefully. Particular attention must be paid to keep combustible structure members away from these chimneys.

Side wall venting of new high efficiency units is another option. This can save a considerable amount of money on the original installation. The boiler or furnace may be equipped with a draft inducer or forced draft burner to make this type of venting possible. It should be noted that releasing oil burner flue gas through the side wall of a building can have some negative results. The flue gas will stain and possibly damage the

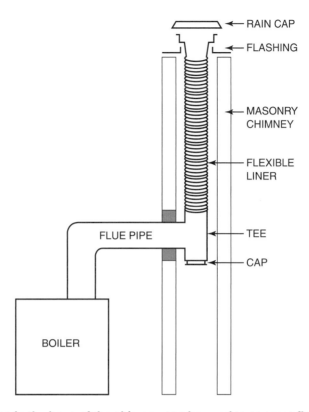

Figure 4-34   **A tee in the base of the chimney can be used to connect flexible liner.**

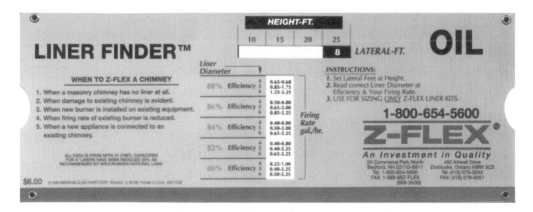

Figure 4-35   **Slide rule to determine chimney liner size.**   *Courtesy: Z-FLEX.*

siding. The odor of the gas is not pleasant and will cause customer complaints and unnecessary service calls. Some side wall venting systems include a draft inducer and an outside exhaust terminal (see Figures 4-36 and 4-37). The direct exhaust system shown in Figure 4-38 includes an enclosed fresh air duct to the oil burner air intake opening. This duct must include a vacuum breaker—usually a draft regulator—so that internal building air will be available to the burner if the outside air intake opening is blocked. This system requires either a forced draft burner or draft inducer on the exhaust to the outside of the building.

Figure 4-36   **Draft inducer for a side wall venting system.** *Courtesy: Field Controls.*

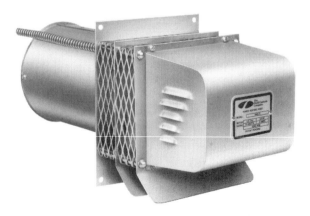

Figure 4-37   **Outside exhaust terminal has fresh air intake openings to supply combustion air to the burner.** *Courtesy: Field Controls.*

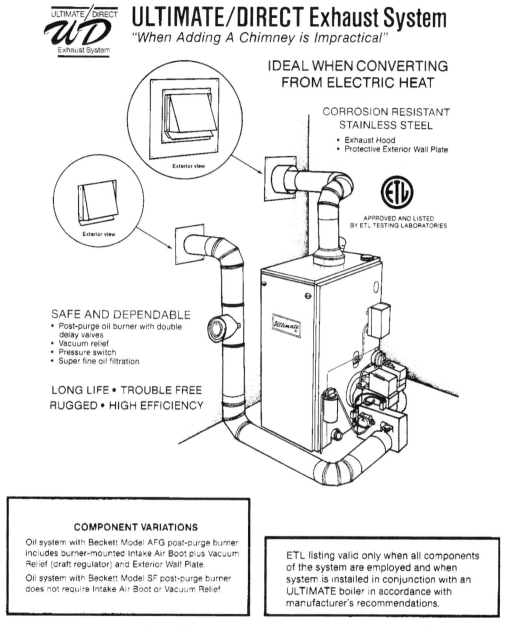

ULTIMATE/DIRECT Exhaust System
"When Adding A Chimney is Impractical"

IDEAL WHEN CONVERTING
FROM ELECTRIC HEAT

CORROSION RESISTANT
STAINLESS STEEL

- Exhaust Hood
- Protective Exterior Wall Plate

APPROVED AND LISTED
BY ETL TESTING LABORATORIES

Exterior view

Exterior view

SAFE AND DEPENDABLE
- Post-purge oil burner with double delay valves
- Vacuum relief
- Pressure switch
- Super fine oil filtration

LONG LIFE • TROUBLE FREE
RUGGED • HIGH EFFICIENCY

COMPONENT VARIATIONS

Oil system with Beckett Model AFG post-purge burner includes burner-mounted Intake Air Boot plus Vacuum Relief (draft regulator) and Exterior Wall Plate.

Oil system with Beckett Model SF post-purge burner does not require Intake Air Boot or Vacuum Relief.

ETL listing valid only when all components of the system are employed and when system is installed in conjunction with an ULTIMATE boiler in accordance with manufacturer's recommendations.

Figure 4-38 **The Ultimate Engineering direct exhaust and air intake system includes a vacuum breaker on the air intake to allow room air to enter the burner if the outdoor terminal is blocked.** *Courtesy: Ultimate Engineering Corp.*

Figure 4-39   **Stack mounted draft inducer.**   *Courtesy: Field Controls.*

DRAFT INDUCERS

When the installation cannot provide the required natural draft, a draft inducer must be installed (see Figure 4-39). A draft inducer is a motor-driven fan that fits into the flue pipe. On a call for heat, the draft inducer starts and establishes the draft in the boiler. Once the draft is established, a control mounted in the flue pipe will feel the draft and allow the burner to start. The addition of a draft inducer will create some additional service problems so it should only be used when absolutely necessary. An alternative to a draft inducer is a forced draft oil burner.

STACK DAMPERS

Another piece of auxiliary equipment, which may be found on some installations, is a stack damper. This is a damper that fits into the flue pipe. The purpose is to eliminate stand-by heat loss from the boiler when the burner is not running. Stand-by heat loss is the result of the chimney draft pulling cold air into the boiler through the oil burner air intake slots. When a stack damper is installed, the sequence of operation is changed so that on a call for burner operation the stack damper must open before the oil burner can start. This obviously adds potential problems to the installation. Stand-by heat loss is not a big problem with oil burner installations that have flame retention head burners or burners equipped with an air intake shutter that closes when the burner shuts off.

SERVICING THE AIR DELIVERY SYSTEM

The Steps for Servicing the Air Delivery System Are:

1. Check the fresh air intake louvre to be certain it is not blocked.
2. Inspect the flue gas passages of the boiler for soot deposits. Clean them if necessary.
3. Check the base of the chimney for dirt. You can check the entire chimney by placing a mirror into the base. You should be able to see blue sky. If you do not, there is an offset or obstruction in the chimney. Remove the obstruction if you can.
4. Another method of checking a chimney is to place a piece of newspaper into the draft regulator. Ignite the paper with a match. The draft of the chimney should pull the flame from the burning paper up into the chimney. In fact, the paper will be pulled into the chimney if the chimney is in good working condition.
5. You can check the chimney with a draft gauge installed through an opening in the flue breeching. You should get a reading of $-.06''$ to $-.09''$ on the gauge.
6. Check the over-the-fire draft. You should be able to adjust it to $-.02''$ with the draft regulator.
7. The fan and burner chassis should be cleaned at least once each year. Remember, the oil burner breathes the surrounding air. If the area around the burner is dirty that dirt will end up inside the burner, especially in the blades of the fan. Some common problems are dogs or cats that sleep in the boiler room, clothes drying in the boiler room, dirt floors, or just dirty people.
8. Flame retention heads must be checked to be certain there is no carbon buildup in the slots that will restrict the air movement.

To Briefly Summarize

1. A permanent source of fresh air is essential.
2. The chimney and flue gas passages must be kept clean.
3. The chimney must provide the required draft.
4. The parts of the system on the burner must be cleaned annually.
5. The flame retention head must be set accurately.

Please Answer These Questions

1. What effect would a clogged chimney have on the combustion process?
2. How do you check a chimney to see if you are getting the required draft?
3. What are the two types of drafts and how are they created?
4. What is the difference between a flame retention head burner and a forced draft burner?
5. What size would the fresh air louvre be for a burner that is firing 1.65 gallons per hour?

6. What is the air delivery pattern delivered by a:
    A. Shell head oil burner
    B. Flame retention head oil burner
    C. Standard oil burner
7. Explain why 1,540 cubic feet of air is required to burn one gallon of fuel oil.
8. What happens in the fire box of a boiler?
9. What effect would clogged flue gas passages have on the operation of the oil burner?
10. When should a draft inducer be installed?

CHAPTER

5

THE FUEL DELIVERY SYSTEM—OIL TANK INSTALLATIONS

The fuel delivery system provides a steady supply of atomized fuel oil, which is mixed with the air supplied by the air delivery system for the combustion process. In order to accomplish this task, the oil must be stored at the building so that it is available when there is a call for heat. Fuel oil storage tanks may be installed inside or outside the building, either buried or aboveground, depending on the local building code restrictions. When the location for the tank is being considered, keep in mind the flash point and the pour point of the fuel in your area. Placing the tank outside on the south side of the building or in a hot boiler room can lead to dangerous vapors as well as loss of fuel by evaporation. Serious service problems can also result from cold storage of fuel oil. Extreme cold causes the formation of wax crystals that will clog filters and oil lines. Fuel oil dealers and their service departments often inherit these problems. Solutions vary from switching to kerosene to avoid wax formation in a cold tank to building a shed to provide shade for that hot tank on Route 52. The tank hook-up becomes a part of the oil burner system and the service technician must be thoroughly familiar with each type of installation.

Oil tanks are made in a variety of sizes, starting with the 275, 550, 1,080, 1,500, 2,000, 2,500, 3,000, and 5,000 gallon tanks. The tanks in your area may vary in size. Generally the 275-gallon tanks are used in gravity feed systems while the larger tanks must be connected in a two-pipe system.

FUEL STORAGE TANKS—GRAVITY SYSTEMS

Tanks that hold 275 gallons are approximately 72″ long, 42″ high and 26″ wide. There are some variations that include a round tank that is approximately 30″ in diameter and 90″ long and a standard tank that is set into the building flat instead of standing upright. Another variation is a tank that stands vertically and can be placed in a closet. These tanks must be approved by the local building department before they are installed. All the 275-gallon tanks have four 2-inch pipe tappings on the top and a 1/2″ pipe tapping on the bottom or on the end near the bottom. There are four brackets welded along the bottom where pipe legs can be attached to the tank. The legs are usually cut 10–12″ long to keep the bottom of the tank off the ground. The legs should be adjusted to allow the tank to pitch toward the bottom 1/2″ tapping. This will allow small quantities of water due to condensation in the tank to run through to the burner where it can be mixed with oil and eliminated without creating a problem with the burner. By continuously getting rid of small quantities of water, we prevent the build up of water that can affect the tank as well as the operation of the burner.

These Tanks Are Connected as Follows (see Figure 5-1):

1. The first tapping, at the end of the tank opposite the bottom tapping, is for the fill line. This is made of 2″ black pipe and starts at the tank and ends with the approved fill box. The oil truck driver connects the delivery hose to the fill box when making a delivery. Local building codes may include the type of fill box and the location on the building or in the sidewalk where it can be installed. Swing joints, a combination of two elbows or an elbow and 45 connected by a short pipe nipple, are used to provide pitch so that no oil will remain in the fill pipe but will drain into the tank. Many outside tank installations have a fill connection right at the tank. A good grade of oil line pipe joint compound must be used on the male threads to prevent leaks.

2. The next tapping is for the vent line. The vent line allows air to enter the tank so that the oil can run out. This line starts at the tank with the Ventalarm, a device that allows the oil truck driver to deliver oil without going into the building to check the amount of oil in the tank first (see Figure 5-2). The Ventalarm has a tube with a whistle built into it that extends about 4″ into the tank. As the tank is filled, air moves through the tube creating a loud whistling sound that can be heard by the driver. When

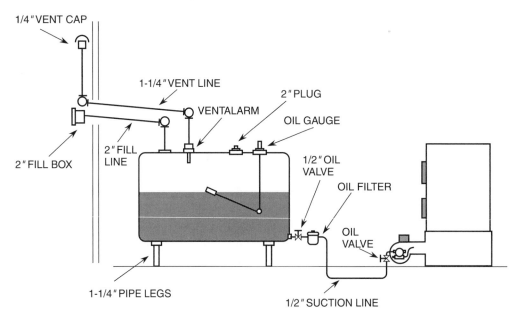

Figure 5-1   **Gravity feed system with 275-gallon oil storage tank.**

Figure 5-2 **Scully Ventalarm signal allows the oil truck driver to fill the oil tank without checking to see how much oil is in the tank.** *Courtesy: Scully Signal Company.*

the level of oil in the tank reaches the bottom of the tube, the whistling stops and the driver knows that the tank is almost full. He can then even off the gallonage number on the meter and stop pumping. The vent line continues, using 1-1/4″ black pipe, outside the building where it is attached to a wall. The vent line must end with an approved vent cap. Local building codes may include the type of cap that is acceptable and where it should be located. It is suggested that the vent cap always be at least two feet above the fill box. Swing joints are used to provide pitch so that any oil that gets into the vent line will drain into the tank.

3. The next tapping is plugged. This opening is used when the oil in the tank has to be measured with a ruler or when water gets into the tank and has to be pumped out.
4. The last tapping is for the oil gauge that will indicate how much oil is in the tank. These are inexpensive float operated gauges that read in 1/8ths of a tank (see Figure 5-3).
5. The 1/2″ tapping at the bottom is for the oil line that runs to the burner. A short nipple and valve are installed, followed by a filter that has a replaceable cartridge. This is the best place to install the filter because it will keep the oil line to the burner clean.
6. The oil line to the burner starts from the filter. This line is usually run in 1/2 or 3/8″ copper tubing that must be buried or covered with concrete to protect it. Some building codes require a protective covering on buried copper tubing. Copper tubing is used because it comes in 30 or 60′ rolls and can be connected with no joints. Flare-type connectors should be used for oil line piping. The oil line is connected to the fuel pump at the burner. A valve is installed at the pump so that the flow of oil can be controlled when the burner is serviced.

It is possible in some areas to install two 275-gallon tanks in the same building. The advantage to this is that by doubling the storage capacity you can increase the size of

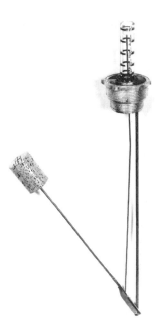

Figure 5-3 **Float-operated oil storage tank gauge for 275-gallon tanks.** *Courtesy: Scully Signal Company.*

each delivery and reduce the number of deliveries to the building during the year. This is good for both the customer and the dealer and should be done if the local code will allow it. The tanks should be set up next to each other and they should line up so that the tops of the tanks are the same height.

The Hook-Up Is Done as Follows (see Figure 5-4):

1. The fill line is connected to the end tapping, on the end of one of the tanks opposite the 1/2″ tapping at the bottom.
2. Two of the other tappings on top of the tank are plugged tight.
3. The last tapping is connected to the second tank by a spill-over line made of 2″ pipe and fittings. Swing joints should be used to prevent leaks or strain on the piping if one of the tanks should settle.
4. The second tank gets the vent line with the Ventalarm. The most convenient top tapping can be used for this line.
5. The next tapping should be plugged.
6. The final tapping on the second tank gets the oil gauge.
7. The tanks are connected together on the bottom, with each tank getting its own oil valve. The filter is installed after the connection between the tanks.
8. The oil line is run to the burner just as in the single tank installation.

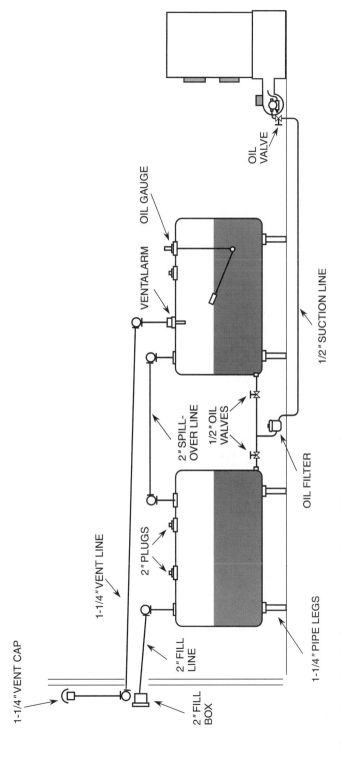

1-1/4" VENT CAP

1-1/4" VENT LINE

2" FILL LINE

2" FILL BOX

2" PLUGS

2" SPILL-OVER LINE

VENTALARM

OIL GAUGE

1/2" OIL VALVES

OIL FILTER

1-1/4" PIPE LEGS

1/2" SUCTION LINE

OIL VALVE

Figure 5-4 **Gravity feed system with two 275-gallon oil storage tanks.**

The tanks are filled by completely flooding the first tank. As the oil pours into the first tank, the air in the tank passes through the spill-over line into the second tank. Air is forced from the tank through the Ventalarm and into the vent line. When the first tank is filled, oil passes through the spill-over line into the second tank. The driver will continue to pump oil until the Ventalarm stops whistling, and then he will stop pumping. It is obvious that the second tank is not completely full and because the tanks are connected at the bottom, the tanks will eventually level off. The oil gauge will indicate the level of oil in the second tank that is matched by the oil level in the first tank. It is not possible to have a gauge in the first tank because it would leak when the tanks are being filled. It is extremely important to have all the plugs tightened in place so that there are no leaks during filling.

The single or double tank installation just described will feed oil to the burner by gravity flow. This is because the tanks usually stand on the basement floor and are raised above the level of the burner by the legs. This means that the fuel pump does not have to create a vacuum to move the oil from the tank to the burner. The fuel pump is set for single pipe. The internal bypass is left open so that the excess oil pumped that does not go through the nozzle will recirculate inside the pump. This creates a problem because the oil movement from the tank will match the nozzle size and this slow movement will allow dirt or wax build-up in the oil line, even though there is a filter in the line.

It is possible to install these tanks below the oil burner, which will prevent the gravity flow of oil from the tank to the burner. Even if the tank is partially below the burner, the gravity flow will be affected. It is still possible to operate this type of installation with a single suction line connected to the bottom of the tank or by running the suction line down into the tank from the top by using a double tapped bushing or slip through connector. Any time the pump must perform the suction function (create a vacuum to move the oil from the tank to the burner) the installation of choice should be the two-pipe system used with larger tanks.

Oil Safety Valve

As gravity tank installations age, there is a serious potential for oil leaks in the buried oil line between the tank and the burner. This type of leak is extremely difficult to detect because the oil will drain into the ground and might not stain the surface where it would be discovered. Installing an oil safety valve, manufactured by the Webster Heating Products, Inc. will eliminate the problem (see Figure 5-5). The oil safety valve (OSV) is a diaphragm-operated, spring-loaded valve that requires approximately 3 inches of vacuum to open. The vacuum is supplied by the fuel pump, even though it is still connected single pipe with no return to the tank. The OSV is installed at the tank, as close to the outlet of the oil filter as possible (see Figure 5-6). When the burner is off, the spring-loaded valve is securely closed and keeps the oil where it belongs, in the tank. When the burner starts, the pump will create the vacuum necessary to open the valve so that oil will flow to the burner. If there is a leak in the oil line, which is now a true suction line, it is quite possible that the pump will draw air into the line and prevent the vacuum necessary to open the

Figure 5-5 **Oil safety valve recommended for gravity feed oil line systems to stop oil flow if the oil line leaks.** *Courtesy: Webster Heating Products, Inc.*

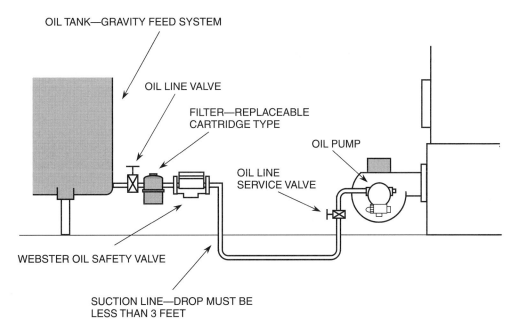

Figure 5-6 **Piping hook-up for the Webster oil safety valve.** *Courtesy: Webster Heating Products, Inc.*

OSV. This will result in a safety lockout of the burner because there will be no oil for the flame. The installation instructions must be followed carefully. The vertical drop from the tank to the lowest point in the suction line must not be more than three feet. This will prevent the siphoning action of the oil dropping through the line from opening the valve.

Another type of safety oil valve is the Firomatic valve. These valves have fusible handles that will melt if there is a fire near the burner. Once the handle melts, a spring inside the valve will close the valve, stopping the flow of oil from the tank. It is absolutely necessary for oil burner technicians to be thoroughly familiar with all local codes regarding the installation of oil storage tanks as well as oil burners and boilers.

FUEL OIL STORAGE TANKS—TWO-PIPE SYSTEMS: BURIED TANK INSTALLATIONS

The larger fuel storage tanks are completely different from the 275-gallon tanks. They generally have two sets of three tappings, usually 2″, on the top of the tank and no tappings at the bottom. There are some variations that require using a single tank tapping for more than one connection. They can never be used in a gravity feed system and must be connected by using a two-pipe system. This means that the internal bypass plug must be installed in the fuel pump and that separate suction and return lines must be connected between the tank and the burner. These tanks can be set up on the basement floor or buried in the ground, either inside or outside the building. We will discuss the buried tank installation first.

The Buried Tank Is Installed as Follows (see Figure 5-7):

1. The 2″ fill line is connected at one end of the tank. The piping from the tank to the fill box is made using double swing joints so that the tank can settle without breaking off the fill pipe. The fill box will be set into the sidewalk or lawn area that is accessible to the oil truck driver.
2. The vent line starts with the Ventalarm and the line runs to the vent cap that is against the outside wall of the building. The vent cap should be at least 2′ above the fill box and at least 2′ from any building opening. Check your local building code requirements for the location of both the fill and vent lines.
3. The third tapping in this set is for the stick well. This is a 2″ line that runs straight down to the tank and has a regular fill box at the ground surface level. This line is used to allow us to measure the oil in the tank and to pump water out of the tank if necessary.
4. The next tapping is for the oil gauge tank assembly. This consists of a special bushing that fits into the 2″ tapping that allows us to screw a 1″ pipe into the bottom so the pipe extends into the tank. Inside this 1″ pipe we slide a copper tube that has a sleeve on the end. The sleeve actually touches the bottom of the tank. The copper line is then run to the oil gauge and connected so that it will not leak. The oil gauge has

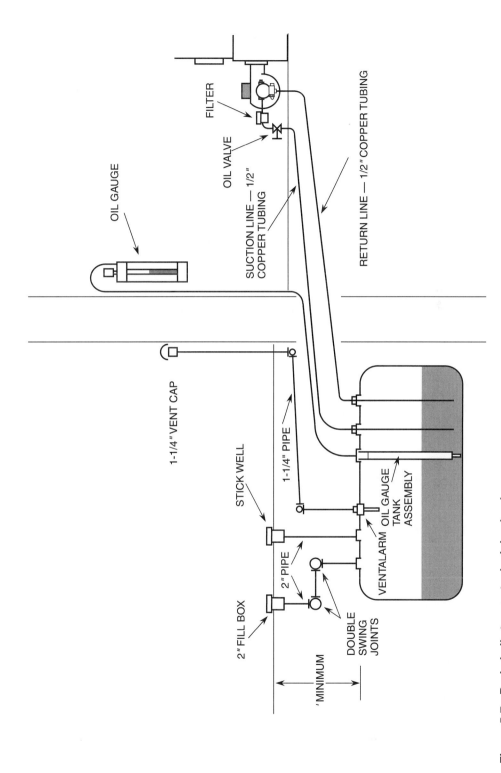

Figure 5-7 **Buried oil storage tank piping hook-up.**

91

a small hand pump on top. To read the depth of oil in the tank you must pump the gauge to force air through the copper line so that there is no oil left in it. Once this is done, the pump is released. As the oil in the tank fills the copper line, air is pushed back into the gauge. This air will move the gauge indicator fluid to the same level as the oil in the tank. Most of these gauges have a chart that will convert the inches shown on the gauge to gallons.

5. The next tapping is used for the suction line. The best method of installing this line is to use a bushing and a slip-through copper connector. This allows us to slide the tubing into the tank so that it terminates approximately 6″ from the bottom. By using the slip-through connector you can make a connection that will not leak. It is absolutely necessary that the suction line be air tight. The copper line is then run to the burner where an oil valve and filter are installed at the fuel pump. Sharp bends in the tubing must be avoided. Clean fill, with no rocks, should be used to cover the oil lines. A protective covering on buried copper tubing is required by some building codes.

6. The last tapping is for the return line and it should be installed exactly the same way we installed the suction line. The return line is connected to the return port of the pump and the end in the tank should terminate 6″ from the bottom of the tank. This is done so that if the suction line develops an air leak we could try to correct the problem by switching the suction and return lines at the burner. No valves are necessary on the return line. It is most important that the internal bypass plug is installed in the fuel pump for this type of system.

The number and size of the tank tappings available in your area may be different from those previously described. It is possible to connect the suction and return line through a single 2″ tapping by using a duplex tank bushing (see Figure 5-8). The fill line can be connected to the same tapping by installing a tee and then using the top of the tee for the suction and return and connecting the fill line to the side outlet of the tee (see Figure 5-9). Regardless of how the tank is connected, use a good quality oil line pipe joint compound—not Teflon tape—on all of the male threads. Pay particular attention to the fill and suction and return lines because of the potential leak hazard.

Figure 5-8 **Duplex tank bushing for suction and return lines.**

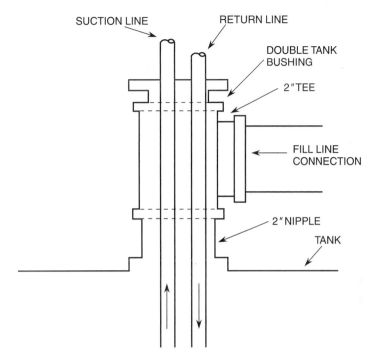

Figure 5-9 **Method of connecting the fill line, suction and return lines to a single 2″ tank tapping.**

Servicing Buried Tanks

Two-pipe systems—particularly buried tanks—seem to be among the more serious problems as far as service work is concerned. It is not unusual for problems with this type of installation to require many visits by service technicians before a solution is worked out. This is quite frustrating for the customer and the company involved and can lead to the loss of an account. The explanation offered below should help to eliminate some if not all of these problems.

The question is, "How does the pump move the oil from a buried tank to the oil burner?" This is the answer (see Figure 5-10). The vent line connected to the oil tank allows air to enter the tank. The air sits on top of the fuel oil in the tank. The weight of the air is known as atmospheric pressure. At sea level atmospheric pressure is equal to 14.7 pounds per square inch of surface area. When the burner runs, the pump gears remove the air from the suction line. As air is removed from the line, a partial vacuum is formed, which simply means that we are reducing the pressure inside the line below 14.7 psi. Reducing the pressure inside the line allows the surface pressure to push the oil into the suction line and start it on its way to the burner. The air is removed from the system through the bleeder port on the pump. This should be left open until a steady stream of oil comes out.

A vacuum gauge must be used to check the operation of the two-pipe system. Install the gauge in the spare suction port of the pump or in a tee that can be installed on

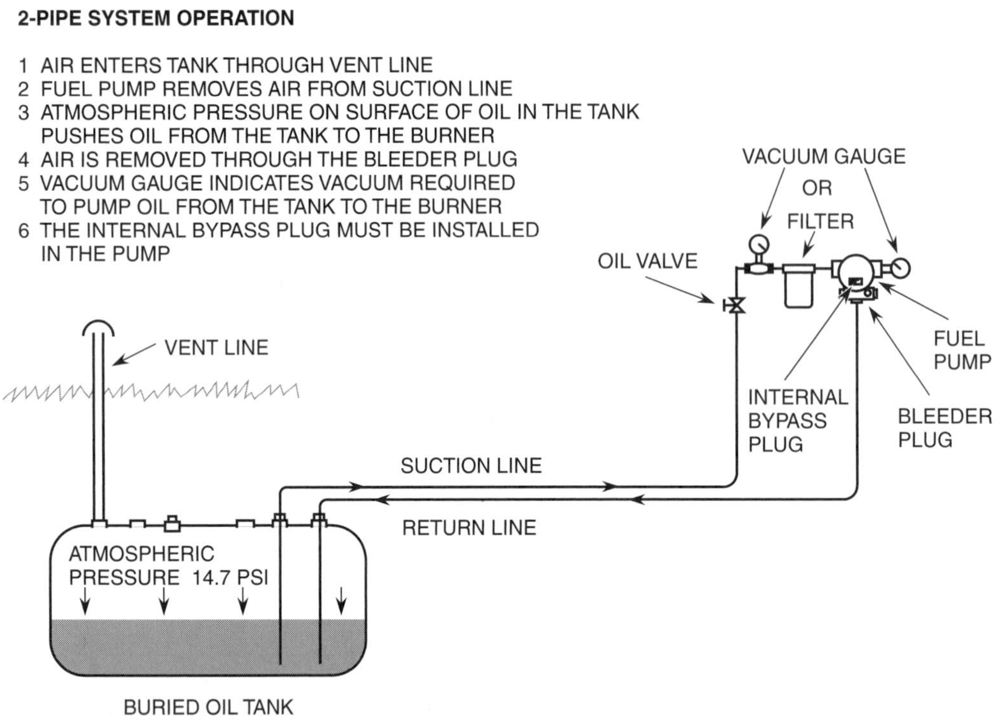

2-PIPE SYSTEM OPERATION

1 AIR ENTERS TANK THROUGH VENT LINE
2 FUEL PUMP REMOVES AIR FROM SUCTION LINE
3 ATMOSPHERIC PRESSURE ON SURFACE OF OIL IN THE TANK
   PUSHES OIL FROM THE TANK TO THE BURNER
4 AIR IS REMOVED THROUGH THE BLEEDER PLUG
5 VACUUM GAUGE INDICATES VACUUM REQUIRED
   TO PUMP OIL FROM THE TANK TO THE BURNER
6 THE INTERNAL BYPASS PLUG MUST BE INSTALLED
   IN THE PUMP

Figure 5-10 **Two-pipe, suction and return system operation.**

the suction line between the oil valve and the pump. The gauge should be installed before the priming process begins. As you bleed the air out of the system, you will see the gauge move up from the 0 mark. This will indicate that you are indeed creating a vacuum inside the suction line. If the gauge does not move, or bounces up and then comes down again, you can be sure that there is a suction leak, which means that air is getting into the suction line somewhere. If this is true you must find and repair the leak in order to get the pump to operate. It is not a good idea to run the pump for any length of time without oil because this can damage the gears. Lightweight lubricating oil or fuel oil should be inserted into the pump through the tapping used for the vacuum gauge before running the pump for the first time.

The vacuum gauge reads in inches of mercury. You can see that the gauge reads from 0 to 30″. You can convert the inches of vacuum to psi by dividing the reading on the gauge by 2. A vacuum reading of 12″ indicates that 6 psi is pushing the oil from the tank to the burner. The vacuum required to move the oil from the tank to the burner is different on every installation.

Factors Involved in Determining the Running Vacuum:

1. How deep is the tank buried or how high do you have to lift the oil? Every foot of vertical lift will require approximately 1″ of vacuum on the gauge.
2. How far is the tank from the burner? Every 10′ of horizontal run will require approximately 1″ of vacuum on the gauge.
3. What kind of pipe was used for the suction line? We recommend the use of 1/2″ copper tubing for several reasons. The line can be installed in one piece with no joints that can be potential suction leaks. The bends and turns can be made gradually, reducing the piping friction loss, which would add to the required vacuum. The copper tubing is not subject to rust and will not rot out.
4. Another factor is the viscosity of the oil. In areas where the temperature remains below freezing for long periods of time, the tank and lines should be buried below the frost line. As the temperature of oil is reduced, the viscosity increases, which will make pumping difficult. Remember that the pour point temperature of # 2 fuel oil is 20°F. At this temperature, wax crystals form that can block the suction line and make pumping impossible. A possible solution would be to switch to kerosene to get through the severely cold months.

Suction line piping must be airtight because if air leaks into the line there will be no reduction in the internal pressure that allows the atmospheric pressure inside the tank to push the oil through the line. A good grade of pipe joint compound (not Teflon tape) should be used. The fuel pump manufacturers will not honor the warranty on their pumps if Teflon tape is used. The reason for this is that if pieces of the tape get into the pumping gears, the gears will be damaged and not perform as expected.

A properly designed two-pipe system should operate at a running vacuum below 12″. Running at higher vacuums may result in the formation of vapor pockets inside the oil line. This is due to the lowering of the vaporizing temperature of the oil because of the high vacuum. This is very possible when the oil lines are run next to hot piping. As a service technician, you will inherit some poorly designed systems that can be service problems. Two-stage fuel pumps can help when high running vacuums are encountered that are the result of the tank being buried too deep or set too far from the burner. A two-stage pump has two sets of pumping gears. One set is used to create the required vacuum and deliver the oil to the second set of gears that will then pressurize the oil and send it to the nozzle. The internal bypass plug *must* be installed in two-stage pumps to convert them to two-pipe operation. High running vacuum can also be caused by partially blocked suction lines because of dirt or valves that are not opening completely. The pump strainer and filters should be checked for dirt. Please do not forget to replace the cover gaskets on the pump or filter, because if air leaks in through either cover the oil will not move out of the tank. Another potential air leak may be the valve packing nut. Make certain to check to see if it is tight. On older installations there may be a foot valve in the tank. This is a ball check valve that used to be installed on the end of the suction line to prevent the draining of the suction line when the burner is shut off. If a foot valve is stuck the only thing you can do is dig down to the tank and replace the suction line.

Buried tanks present many problems to the service technicians because of their age. Over a long period of time water can collect in the bottom of the tank. Water can get into the tank through the fill box cover, the stick well cover, or any broken underground piping. If the tank is not filled during the summer, condensation will form on the inside of the tank. Since there is no way for the water to get out of the tank, condensation can build up to a considerable amount of water over a period of years. Since water is heavier than oil, it will collect on the bottom of the tank and just sit there. Over a period of years a layer of sludge will build up on the bottom of the tank. Sludge is made up of water, dirt that was in the oil, a wax that the water forms when it contacts the oil, and algae, which is a form of plant life that grows in the sludge. The sludge is an acid that attacks the steel the tank is made of and will eventually create a leak. The sludge layer does not become a service problem until it gets thick enough to reach the bottom of the suction line. Once that point is reached, sludge will get into the filter, the pump strainer, and the nozzle, resulting in a burner shut down. A floating suction line can be installed in tanks that are having serious sludge problems. The end of the suction line is attached to a float that keeps clean oil flowing off the top of the oil in the tank (see Figure 5-11).

You can determine the level of water in a buried tank by using a ruler that has water testing paste on it. Put the paste on at least 8–10″ of the ruler and insert the ruler into the tank through the stick well. The indicator paste is blue and will turn red if there is water in the tank. Any significant amount of water should be removed before it creates problems. You can remove the water by pumping it out through the stick well. Drop a piece

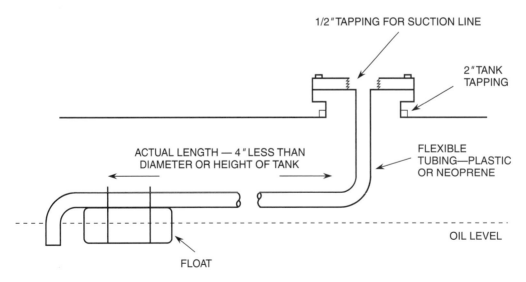

Figure 5-11 **Floating suction line used when tank is contaminated. This will draw clean oil from the top of the tank.**

of pipe or copper tubing into the tank until it hits the bottom. Connect the pipe to a pump set and pump the water out. Water coming out of the tank will be milky white in appearance. You will see a distinct change in the color of the liquid when you start pumping oil. It is impossible to remove all the water from the tank by this method but you will lower the water (and sludge level) below the end of the suction line to prevent further burner problems. Many local communities require the testing of buried tanks for leaks. Your company may be involved in performing the testing and even replacement of buried tanks. Try to remember to avoid building service problems into the tank installations.

FUEL OIL STORAGE TANKS—TWO-PIPE SYSTEMS: SURFACE INSTALLATIONS

Large tanks that are installed on the surface have some problems that are unique to them. They can be protected by being covered with concrete, concrete or cinder blocks, or by being placed inside a vault room made of concrete blocks. Please check your local building code requirements when installing this type of tank. When the tanks are vaulted, the wall of the vault must have weep holes every two feet along the bottom. These openings are there to let us know if the tank is leaking.

The piping hook-up is almost identical to that of the buried tanks (see Figure 5-12). The difference is in the suction and return lines. Because the burner is usually below the tank, and the suction line is connected to the top of the tank, we must provide some method of controlling a siphon. A siphon is created by the oil dropping through the suction line to the burner. The weight of the oil causes a reduction in the internal pressure of the suction line, which allows the atmospheric pressure on the surface of the oil to push the oil through the line. Once the air is out of the line, oil will siphon from the tank to the burner as easily as in a gravity feed system. The danger here is that if a leak develops near the fuel pump, it would be possible to siphon all the oil out of the tank onto the basement floor. To prevent this, an antisiphon valve is installed at the highest point in the suction line.

The antisiphon valve requires the vacuum created by the pump to open it. Once the burner shuts off, and the vacuum drops to 0, the antisiphon valve closes to keep the oil in the tank from continuing to run through to the burner. These valves add about 4″ to the running vacuum but are not a problem as long as they stay clean and do not get stuck closed. A high vacuum reading on this type of installation might very well be caused by a stuck antisiphon valve. Antisiphon protection may be required on a buried tank installation if the tank is above the level of the oil burner.

The return line must terminate at the top of the tank instead of being run down to the same level as the suction line. This is to prevent a siphon of oil if a leak should develop on the return line. A check valve is installed on the return line near the burner with a union between the check valve and the pump. This is done to make it possible to replace the pump without having the oil in the return line run out during a pump replacement. A regular hand valve should not be used because if it were to be shut off during normal operation, you

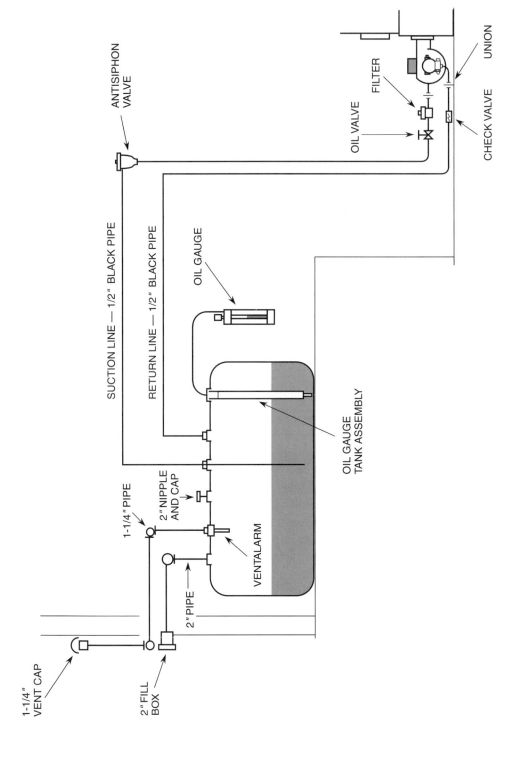

ANTISIPHON VALVE

FILTER

UNION

OIL VALVE

CHECK VALVE

SUCTION LINE — 1/2" BLACK PIPE

RETURN LINE — 1/2" BLACK PIPE

OIL GAUGE

OIL GAUGE TANK ASSEMBLY

1-1/4" PIPE

2" NIPPLE AND CAP

VENTALARM

2" PIPE

1-1/4" VENT CAP

2" FILL BOX

Figure 5-12 **Two-pipe hook-up when the tank is level with or above the oil burner.**

would blow the pump shaft seal out and flood the burner with oil. It is absolutely essential to install the internal bypass plug in the pump on these installations.

The running vacuum should not exceed 12″. One of the problems encountered with high running vacuums is the creation of vapor pockets in the suction line. Fuel oil will start to vaporize at its flash point at atmospheric pressure. As the pressure on the oil is decreased by being in a partial vacuum, the temperature of vaporization will also decrease. If the oil line is running alongside a hot return pipe, it is easy to see that a vapor pocket could form. The result would be much the same as having air in the line and the pump would have to be primed to remove the vapor. This could lead to a safety lockout of the burner and a no heat service call. It is difficult to diagnose this problem because by the time the service technician gets to the job, the condition no longer exists. Only by using a vacuum gauge can you be certain to cover all the possible problems that a two-pipe installation presents.

Remember these simple rules:

1. A low or fluctuating vacuum reading with an unsteady flow of oil is a sure indication of a suction leak.
2. A high vacuum reading and possibly a screaming pump, is a sure indication of a clogged suction line.
3. Every installation has a built-in running vacuum that you, as a service technician, have no control over.
4. Two-stage fuel pumps should be used on two-pipe installations.
5. The internal bypass plug must be installed in the fuel pump, single- or two-stage, to convert it to a two-pipe system.

To Briefly Summarize

1. All fuel oil storage tank installations must comply with the local building code requirements.
2. 275-gallon tanks, or tanks of that approximate size, can be connected to the burner to provide gravity feed of fuel. Some localities allow two of these tanks in the same building.
3. Larger tanks require a two-pipe system and suction and return lines to feed the oil to the burner.
4. The internal bypass plug must be installed in the fuel pump to operate in a two-pipe system.
5. Two-stage fuel pumps are recommended for two-pipe installations.

Please Answer These Questions

1. How can you determine if there is water in a buried oil tank?
2. What is an antisiphon valve and when is it used?
3. Why is the Ventalarm an important part of every oil tank installation?

4. Why are the suction and return lines on a buried tank both run to within 6″ of the bottom of the tank?

5. Why can't the return line run to the bottom of a tank that is installed level with or above the oil burner?

6. What is a two-stage pump and how does it operate?

7. What does the term *gravity feed* mean?

8. How much oil runs through the suction line of a gravity feed system?

9. Explain how the fuel pump moves the oil from a buried tank to the oil burner.

10. How do you prepare a fuel pump for installation in a two-pipe system?

6

THE FUEL DELIVERY SYSTEM—PUMPS AND NOZZLES

Once the oil is in the storage tank, the oil burner is responsible for moving the oil from the tank to the burner and preparing it for the combustion process. To accomplish these tasks, the oil burner has the following parts:

1. The motor
2. The oil pump coupling
3. The fuel unit—commonly referred to as the *pump*
4. The oil burner nozzle
5. The nozzle adaptor
6. The delayed opening oil valve
7. The copper tubing connection between the fuel unit and the nozzle assembly

FUEL SYSTEM OPERATION

The motor drives the oil burner. The shaft of the motor and the shaft of the fuel unit line up perfectly when these parts are mounted on the burner chassis. The fan is attached to the motor shaft so the closed end of the fan is very close to the face of the motor (see Figure 6-1). This leaves room on the motor shaft for the oil pump coupling. The oil pump coupling has a rubber shaft and two metal or plastic ends that fit on the motor and fuel unit shafts. When the motor is running, the fan forces air through the burner and the fuel unit forces fuel oil through the nozzle.

FUEL UNITS (PUMPS)

The Fuel Unit, or Pump, Has These Functions to Perform:

1. Move the fuel from the storage tank to the burner.
2. Filter the oil.

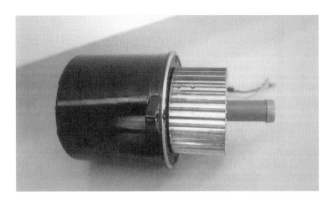

Figure 6-1 **Oil burner motor, fan, and oil pump coupling.** *Courtesy: R. W. Beckett Corp.*

3. Pressurize the oil so that it will atomize as it leaves the nozzle.
4. Provide clean, sharp cutoff of the oil at the end of the burner cycle.

The fuel unit itself is a carefully constructed, compact machine. It has a cast-iron body with a number of openings or ports for connection to the storage tank and the nozzle. Each of these ports must be identified and properly connected for the unit to operate properly (see Figures 6-2 and 6-3). The suction ports lead directly into a large chamber that contains the pump strainer. These strainers are fine mesh screens made of steel or

Figure 6-2 **Suntec Model A fuel unit, commonly referred to in the trade as the pump.** *Courtesy: Suntec Industries, Inc.*

Figure 6-3 **Webster Model M two-stage fuel unit.** *Courtesy: Webster Heating Products, Inc.*

plastic. The Webster fuel unit employs a blade or rotary filter attached to the pump shaft that cuts or grinds up the dirt instead of a strainer (see Figure 6-4). The function of the screen is to catch dirt before it can get into the pumping gears and damage them. The pumping gears fit directly under the strainer. Oil flows into the gears, which then squeeze the oil through a passage to the pressure regulator. The pressure regulator is a piston that has a heavy spring inserted into one end and a neoprene disc attached to the other end (see Figure 6-5). The neoprene end fits tightly against a seat that leads to the nozzle port of the pump. The spring forces the piston against the seat sealing it. When the motor is running, the pump shaft turns and turns the pumping gears. The gears force oil into the pressure regulator, which forces the piston back against the spring. When the pumping pressure reaches 100 psi, the spring is compressed and the piston moves away from the seat allowing oil to flow through the nozzle port to the nozzle. The pumping gears can pump more oil than will flow through the nozzle. The excess oil goes through an internal bypass in the pump and back into the strainer chamber where it goes through the strainer back to the gears and around again. When the motor stops, the spring forces the piston into the closed position and cuts off the oil to the nozzle (see Figure 6-6).

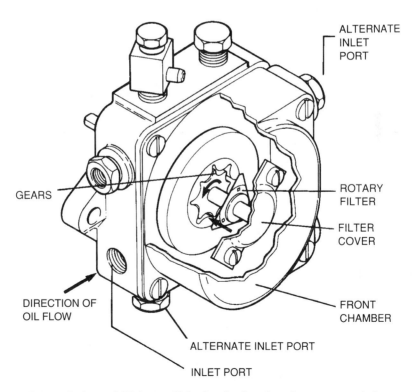

Figure 6-4 **Internal view of Webster M fuel unit showing the gears and the rotary filter.**
Courtesy: Webster Heating Products, Inc.

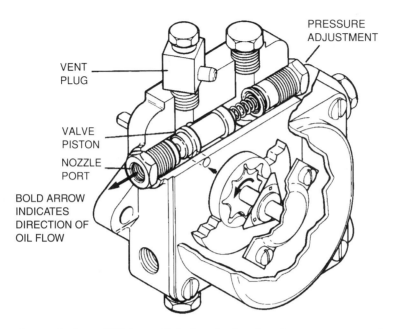

Figure 6-5 **Internal view of Webster M fuel unit showing the pressure regulating valve.**
Courtesy: Webster Heating Products, Inc.

PUMP TESTING PROCEDURES

The normal operating pump pressure for most burners is 100 psi. The pump pressure can be checked by installing a pressure gauge in the port closest to the nozzle outlet port of the pump. The ability of the pump to stop the flow of oil to the nozzle can be tested by installing a pressure gauge directly into the nozzle port. The pumping ability of the pump can be checked by installing a vacuum gauge in one of the suction or inlet ports. These tests are necessary for the service technician to perform when there is a problem with an oil burner that involves the fuel delivery system. The testing procedures are as follows:

The Pressure Test

This Is How to Do the Pressure Test:

1. Install a pressure gauge in the priming port or the port marked as the pressure gauge port of the pump, closest to the nozzle outlet port of the pump. This gauge should be capable of withstanding pressures of at least 300 psi (see Figure 6-7).
2. Run the burner and observe the flame.
3. Raise the pressure by screwing the pressure adjusting screw into the pump. The pressure should rise as you do this. Observe the flame.

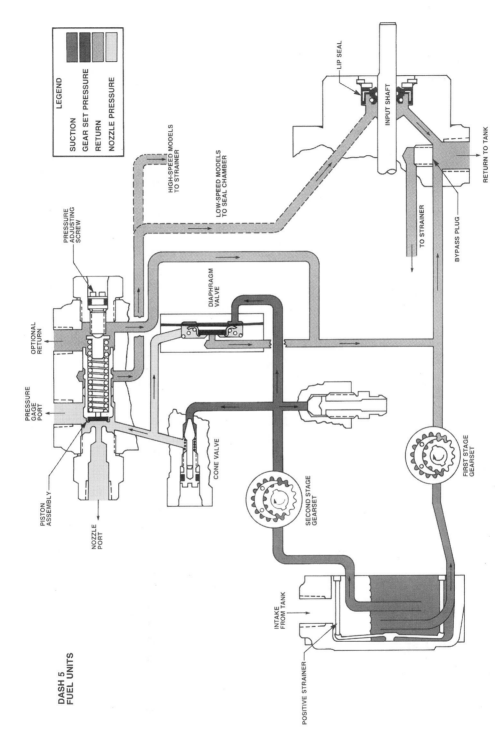

DASH 5 FUEL UNITS

LEGEND

SUCTION
GEAR SET PRESSURE
RETURN
NOZZLE PRESSURE

PRESSURE ADJUSTING SCREW

HIGH-SPEED MODELS TO STRAINER

LOW-SPEED MODELS TO SEAL CHAMBER

LIP SEAL

INPUT SHAFT

RETURN TO TANK

TO STRAINER

BYPASS PLUG

OPTIONAL RETURN

DIAPHRAGM VALVE

PRESSURE GAGE PORT

PISTON ASSEMBLY

CONE VALVE

NOZZLE PORT

FIRST STAGE GEARSET

SECOND STAGE GEARSET

INTAKE FROM TANK

POSITIVE STRAINER

Figure 6-6 **Oil flow through the Suntec Model B two-stage fuel unit.** *Courtesy: Suntec Industries, Inc.*

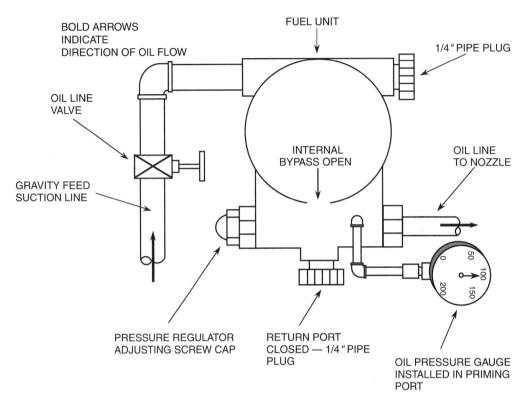

Figure 6-7 **The pressure gauge is installed in the pump port closest to the nozzle outlet for the pump pressure test.**

4. You should be able to raise the pump pressure to 170 psi without losing the flame. The flame should get larger. There should be no loss of fuel to the flame during this test. If the flame gets smaller or cuts off, the piston is sticking and the pump should be replaced.

5. If there is no loss of fuel or flame, the pump is operating properly and you can readjust the pressure to 100 psi. Remove your pressure gauge and plug the port.

The Cutoff Test

This test is done to determine if the pump shuts off the flow of oil when the burner cycle ends. This is important on installations in which the oil is fed by gravity to the oil burner. If the pump does not cut off properly, it is possible for oil to flow from the tank through the pump and nozzle into the combustion chamber. This can lead to the formation of soot, oil smells in the building, and possibly an oil stain around the boiler. If it is not repaired and enough oil leaks around the boiler, it can lead to a fire.

This Is How to Do the Cutoff Test (see Figure 6-8):

1. Disconnect the nozzle line at the pump. Remove the tubing connector and install a pressure gauge.
2. Run the burner—the gauge will read 100 psi.
3. Shut the burner off—the pressure on the gauge will drop to 80 psi and stay there if the pump is seating properly. If the pressure continues to drop, the seat is leaking and the pump must be replaced.
4. If the pump is functioning properly, remove your gauge and replace the fittings and reconnect the nozzle line.

The Vacuum Test

This test is done to determine how well the pump can move the fuel from the tank to the burner. This is necessary when the installation has a large tank, either buried or above ground, that utilizes a suction and return piping system. When this type of system is installed, there must be a plug in the internal bypass port of the pump. This prevents the oil that cannot flow through the nozzle from going back into the filter chamber. The oil must then go back to the tank through a return line that is connected from the return port of the pump (see Figure 6-9).

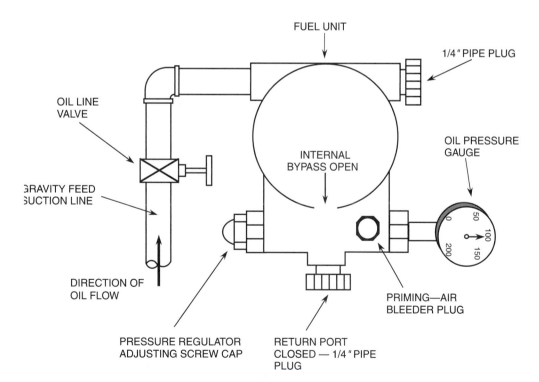

Figure 6-8 **The pressure gauge is installed in the nozzle outlet for the cutoff test.**

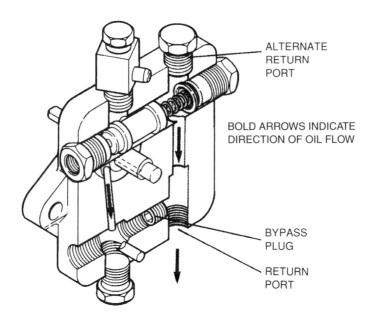

ALTERNATE
RETURN
PORT

BOLD ARROWS INDICATE
DIRECTION OF OIL FLOW

BYPASS
PLUG

RETURN
PORT

Figure 6-9 **Internal view of a Webster M pump showing the internal bypass plug in place to set the pump for two-pipe operation.** *Courtesy: Webster Heating Products, Inc.*

This Is How to Do the Vacuum Test (see Figure 6-10):

1. Install a vacuum gauge in the spare suction port of the pump or in a tee on the suction line.
2. Run the burner. If necessary, prime the pump until you get a steady flow of oil out of the priming plug. If you cannot get the pump primed, or if the flow of oil is not steady, check the gauge reading. A low reading or a fluctuating gauge needle indicates that air is leaking into the suction line. A high reading indicates a clogged suction line. Once the pump is primed, close the bleed valve and check the flame. The flame should be full and steady.
3. Check the vacuum reading on your gauge—it should read less than 12″.
4. Close the valve on the suction line at the burner. The vacuum should jump right up to 25″ or higher. This indicates that the pumping gears are operating properly. If you do not get this kind of result—if the vacuum does not increase rapidly—the pumping gears are probably worn and the pump should be replaced.
5. You can now check the suction line from the tank to the burner. Open the valve at the burner and prime the pump until you get a steady stream of oil coming out. Check the vacuum reading on your gauge. A high reading, above 15″, may indicate a clogged suction line or may be because of the tank being buried too deep or too far from the burner. A low or fluctuating reading on the gauge means that there is a suction leak.

There are no shortcuts in these test procedures. The only way that you can detect problems with the fuel unit is by using the gauges as outlined above. Do not guess and

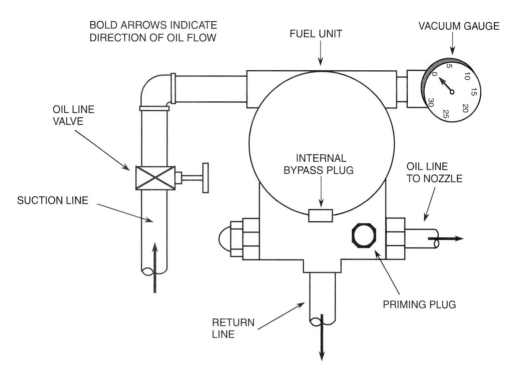

Figure 6-10 **The vacuum gauge is installed in a spare suction port on the pump or a tee in the suction line between the oil line valve and the pump.**

waste time replacing fuel units without testing them first. Nothing can be more embarrassing than replacing a pump and still not being able to get the oil out of the tank. The fuel unit is the heart of the high pressure oil burner. Treat it with the respect it deserves; use gauges to test it!

OIL SUPPLY PROBLEMS

There must be a clean supply of oil to the pump for proper burner operation. On single-pipe gravity feed systems, you can check the flow of oil from the tank to the burner as follows:

1. Close the oil supply valve at the pump.
2. Remove the strainer cover. Place a can under the pump to catch the oil that will spill out when the cover is removed.
3. Remove the strainer. How dirty is it? Are there signs of water in the oil? Is there sludge (a heavy, brown, gooey kind of dirt) in the strainer chamber?
4. Check for gravity flow of oil from the tank by opening the supply valve.

5. Is there a good flow of clean oil from the tank? If not:
   A. Check for a clogged filter cartridge. Close the supply valve before the filter. Remove the cartridge and clean the filter can. Open the supply valve to check the flow of oil into the filter. If the oil flows freely, install a new cartridge and assemble the filter. You must replace the filter cover gasket to prevent an oil leak.
   B. If the oil does not run freely into the filter, attach the filter can without installing a new cartridge. Then either blow the line back to the tank with a CO_2 cartridge or pump the line out with a hand pump.
   C. Do not install a new filter cartridge until there is a steady flow of clean oil at the filter.
6. Once the filter and line are clear, there should be a good gravity flow of oil to the pump. Clean the strainer chamber by allowing oil to run through it with the cover off.
7. Clean or replace the strainer. Be careful that you install a strainer with the appropriate screen size for the burner firing rate. Some strainers require a spring that holds the strainer in place over the gear housing.
8. Replace the strainer cover gasket and bolt the cover to the pump. Tighten all the bolts by hand first then tighten the bolts in an alternating pattern. Do not overtighten the bolts as they do break off. This would then require replacing the pump or drilling and removing the broken bolt.

The strainer must be checked when servicing the pump on two-pipe systems as well. The procedure is quite similar to the gravity feed system except that there will be no oil flow from the tank to help clean the parts. Check for signs of water or sludge in the oil. It may be necessary to clean the tank to remove water or sludge to get the installation to operate properly. All cover gaskets must be replaced to ensure that no air will leak into the suction line.

PUMP REPLACEMENT PROCEDURE

Once a clean supply of oil is available at the pump, the testing procedures can be followed. If the pump fails any one of the tests, it should be replaced. It is pointless to replace a pump without first establishing a clean supply of oil. The failure of the original pump was probably due to contamination in the oil. Just changing the pump will not solve the problem because the replacement will not last very long.

Pump Replacement Requires the Following Information:

1. The manufacturer's name and model number if available.
2. The direction of rotation (looking at the shaft). Does it turn to the right (clockwise–CW)? Does it turn to the left (counter clockwise–CCW)?
3. Where is the nozzle outlet, on the right or left side?
4. How does the pump attach to the burner chassis? Does it have a mounting flange or is it a hub mount?

5. What is the speed of the motor? 1,725 or 3,450 rpm?
6. Is this a single- or two-stage pump? Single-stage pumps have one set of pumping gears (see Figure 6-4) and are generally used for gravity feed systems. Two-stage pumps have two sets of gears (see Figure 6-11); one set supplies the vacuum to get the oil out of the tank and the second set pressurizes the oil. Two-stage pumps are used for two-pipe installations in which high vacuum may be encountered.
7. Is this a gravity feed or two-pipe installation? If it is a two-pipe installation the internal bypass plug must be installed in the appropriate port.
8. Check the shaft size and length. Will you have to change the oil pump coupling?

Obviously an exact replacement is the easiest way to get the burner back into operation. It is possible to switch to another manufacturer's pump if you check out the items listed above to see if the new unit fits and can do the job. In any replacement, you must be certain that the alignment between the motor and pump shafts is perfect and that the burner turns freely.

The Procedure for Replacing the Pump Is as Follows:

1. Disconnect the power to the oil burner.
2. Close the oil line supply valve.

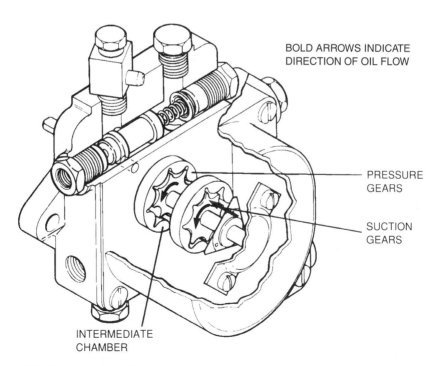

BOLD ARROWS INDICATE DIRECTION OF OIL FLOW

PRESSURE GEARS

SUCTION GEARS

INTERMEDIATE CHAMBER

Figure 6-11 **Internal view of a two-stage Webster M pump showing both sets of gears.**
Courtesy: Webster Heating Products, Inc.

3. Disconnect the oil line or lines. On two-pipe installations, be certain to identify the suction and return lines so that you do not mix them up when making connections to the replacement pump.
4. Remove the connecting fittings from the pump.
5. Loosen the pump coupling set screw on the pump shaft. Many pump coupling ends do not have set screws. The shaft opening in the coupling end has a flat spot that fits against the flat surface on the pump shaft.
6. Unbolt the pump from the burner chassis and slide the pump out of the chassis opening.
7. Slide the replacement pump into the burner. Be certain that the pump coupling engages and that the burner can turn freely. If there is a set screw in the coupling end, tighten it onto the flat surface on the pump shaft.
8. Bolt the pump in place and check for free rotation of the burner again.
9. Install the oil line connecting fittings. Use a good grade of oil line pipe joint compound on the male threads. Do not use Teflon tape on oil line piping. If pieces of the tape get into the pump, they will damage the gears. On two-pipe installations, be certain to install the internal bypass plug.
10. Connect the oil lines.
11. Run the burner and prime the pump. Check for oil leaks.
12. Shut the power off and clean up around the burner. Check for oil leaks and clean any spilled oil from the floor.
13. Run the burner. Check for smooth light off, flame size, smoke, oil leaks, and clean cutoff at the end of the burner cycle.
14. Remove all newspaper, rags, oil absorber, etc. from the premises.

TWO-STAGE PUMPS—TWO-PIPE SYSTEMS

Perhaps this is the time to discuss the confusion concerning two-stage pumps and two-pipe systems. A two-stage pump has two sets of pumping gears attached to the shaft. The first set performs the suction function of the pump, creating the vacuum that can draw oil from the tank. Once the oil arrives inside the pump, the second set of gears pressurizes the oil, forcing it through the pressure regulator to the nozzle. A two-pipe system has a separate suction line to draw oil from the tank to the burner and a return line that carries excess oil back to the tank. The operation of two-pipe systems is explained in detail in Chapter 5, Oil Tank Installations.

It is highly recommended to use two-stage pumps on two-pipe tank installations. You can use a single-stage pump but the two-stage units will perform better, particularly on long suction line installations. Regardless of which type pump you use, you must install the internal bypass plug in the pump to set it up for two-pipe operation.

To Briefly Summarize:

1. The fuel unit, commonly called the pump, is the heart of the high pressure gun type oil burner.
2. There must be a steady supply of clean oil available for proper pump operation.
3. Pump operation must be checked with these tests:
   A. the pressure test
   B. the cutoff test
   C. the vacuum test
4. Single-stage pumps have one set of pumping gears and can be used for gravity feed or low vacuum two-pipe systems.
5. Two-stage pumps have two sets of pumping gears and should be used for two-pipe systems.
6. Two-pipe systems always require the insertion of the internal bypass plug in the pump.

THE OIL BURNER NOZZLE

The importance of the nozzle cannot be overemphasized. This relatively inexpensive part is truly crucial to the proper operation of every oil burner. The manufacturers must be given credit for their efforts, which have resulted in a precision instrument that can deliver an exact amount of fuel in a precise spray pattern and shape designed to match the air delivery pattern of each type of oil burner. This match will allow the burner to mix the air and oil without the necessity for excess air that will reduce the operating efficiency. Service technicians must be able to recognize problems with the nozzle and make appropriate replacements.

Nozzle Functions

The Functions of the Nozzle Are as Follows:

1. The nozzle is the metering device for the high pressure gun type oil burner. The size of the nozzle determines the amount of oil entering the combustion process. Nozzles are rated in gallons per hour at 100 psi of fuel pump pressure. Increasing the pump pressure will increase the amount of fuel passing through the nozzle. Table 6-1 indicates how raising pump pressure affects nozzle capacity.
2. The nozzle must atomize the oil. To *atomize* means to break the oil into tiny droplets that can mix with air and burn. Figure 6-12 shows the droplets of oil leaving the nozzle. Increasing the pump pressure will result in smaller droplets as shown in Figure 6-13.
3. The nozzle delivers the oil droplets in a specific pattern. The delivery patterns are designed to match the air delivery patterns of the oil burners. This is extremely important

NOZZLE CAPACITIES
U.S. Gallons per Hour No. 2 Fuel Oil

rate gph @ 100 psi	Operating Pressure: pounds per square inch							
	125	140	150	175	200	250	275	300
.40	.45	.47	.49	.53	.56	.63	.66	.69
.50	.56	.59	.61	.66	.71	.79	.83	.87
.60	.67	.71	.74	.79	.85	.95	1.00	1.04
.65	.73	.77	.80	.86	.92	1.03	1.08	1.13
.75	.84	.89	.92	.99	1.06	1.19	1.24	1.30
.85	.95	1.01	1.04	1.13	1.20	1.34	1.41	1.47
.90	1.01	1.07	1.10	1.19	1.27	1.42	1.49	1.56
1.00	1.12	1.18	1.23	1.32	1.41	1.58	1.66	1.73
1.10	1.23	1.30	1.35	1.46	1.56	1.74	1.82	1.91
1.20	1.34	1.42	1.47	1.59	1.70	1.90	1.99	2.08
1.25	1.39	1.48	1.53	1.65	1.77	1.98	2.07	2.17
1.35	1.51	1.60	1.65	1.79	1.91	2.14	2.24	2.34
1.50	1.68	1.77	1.84	1.98	2.12	2.37	2.49	2.60
1.65	1.84	1.95	2.02	2.18	2.33	2.61	2.73	2.86
1.75	1.96	2.07	2.14	2.32	2.48	2.77	2.90	3.03
2.00	2.24	2.37	2.45	2.65	2.83	3.16	3.32	3.46
2.25	2.52	2.66	2.76	2.98	3.18	3.56	3.73	3.90
2.50	2.80	2.96	3.06	3.31	3.54	3.95	4.15	4.33
2.75	3.07	3.25	3.37	3.64	3.90	4.35	4.56	4.76
3.00	3.35	3.55	3.67	3.97	4.24	4.74	4.97	5.20
3.25	3.63	3.85	3.98	4.30	4.60	5.14	5.39	5.63
3.50	3.91	4.14	4.29	4.63	4.95	5.53	5.80	6.06
3.75	4.19	4.44	4.59	4.96	5.30	5.93	6.22	6.50
4.00	4.47	4.73	4.90	5.29	5.66	6.32	6.63	6.93
4.50	5.40	5.32	5.51	5.95	6.36	7.11	7.46	7.79
5.00	5.59	5.92	6.12	6.61	7.07	7.91	8.29	8.66
5.50	6.15	6.51	6.74	7.27	7.78	8.70	9.12	9.53
6.00	6.71	7.10	7.35	7.94	8.49	9.49	9.95	10.39
6.50	7.26	7.69	7.96	8.60	9.19	10.28	10.78	11.26
7.00	7.82	8.28	8.57	9.25	9.90	11.07	11.61	12.12
7.50	8.38	8.87	9.19	9.91	10.61	11.86	12.44	12.99
8.00	8.94	9.47	9.80	10.58	11.31	12.65	13.27	13.86
8.50	9.50	10.06	10.41	11.27	12.02	13.44	14.10	14.72
9.00	10.06	10.65	11.02	11.91	12.73	14.23	14.93	15.59
9.50	10.60	11.24	11.64	12.60	13.44	15.02	15.75	16.45
10.00	11.18	11.83	12.25	13.23	14.14	15.81	16.58	17.32
10.50	11.74	12.42	12.86	13.89	14.85	16.60	17.41	18.19
11.00	12.30	13.02	13.47	14.55	15.56	17.39	18.24	19.05
12.00	13.42	14.20	14.70	15.88	16.97	18.97	19.90	20.79

Table 6-1 **Effects of Pressure on Nozzle Flow Rate.** *Courtesy R. W. Beckett Corp.*

Figure 6-12   **Oil droplets spray from nozzle at 100 psi pump pressure.** *Courtesy: Delavan.*

because by matching the nozzle to the air delivery pattern we can get complete combustion with little excess air. Figure 6-14 illustrates the spray patterns available in standard nozzles.

4. The angle of spray as marked on the nozzle will determine the shape of the flame. This is particularly important in the solid air delivery pattern. The flame must be shaped to fit into the combustion chamber without hitting the floor or side walls. Hollow air delivery pattern burners always require a wide angle nozzle. The semi-solid flame pattern burners require a medium angle. Figure 6-15 illustrates the spray angles available in standard nozzles.

Figure 6-13   **Oil droplets spray from nozzle at 300 psi pump pressure.** *Courtesy: Delavan.*

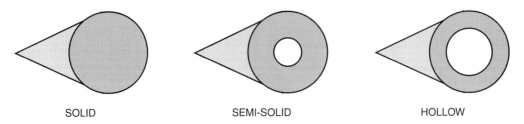

Figure 6-14   Standard oil burner nozzle spray patterns.

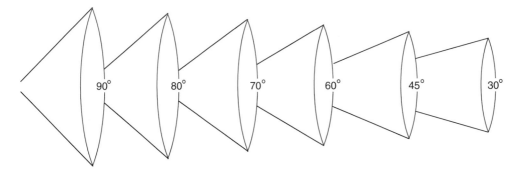

Figure 6-15   Standard oil burner nozzle spray angles.

5. The nozzle strainer filters the oil to keep the interior passages and orifice of the nozzle clean. There are two types of nozzle filters: the wire mesh strainer and the sintered bronze. The wire mesh strainers are used on larger size nozzles while the sintered bronze type is for nozzles in the .50 to 1.00 gallon per hour range. Advances in nozzle technology now include a sintered filter that can be used on nozzles of up to 2.00 GPH firing rates. There are also in-line sintered filters that can be installed in the nozzle line between the pump and nozzle (see Figure 6-16). These filters are fully serviceable and help to eliminate problems on installations in which the oil is contaminated. Another advancement is the small cutoff valve that fits inside the nozzle strainer. This is a spring loaded device that requires 60–80 psi of pump pressure to open. When the burner operating cycle is completed and the burner stops, this valve will close providing clean cutoff of the flame and the elimination of any after drip of oil into the combustion chamber.

Nozzle Operation

Please refer to Figure 6-17, a cutaway view of a nozzle, to help you understand how the nozzle works. The flow rate through the nozzle is determined by the size of the orifice. Nozzles are all rated, in gallons per hour, at 100 psi of fuel pump pressure. The spray angle and pattern are determined by the slots in the distributor, the shape of the swirl chamber, and the orifice. The nozzle filter, either the sintered bronze or fine mesh screen,

Figure 6-16 **Sintered bronze filter is installed between the pump outlet and the nozzle.** *Courtesy: Delavan.*

is designed to keep the interior of the nozzle clean so that the oil can flow through the distributor slots into the swirl chamber and finally out of the orifice with no restrictions.

The energy required to break the oil into tiny droplets is supplied by the fuel pump. The oil is forced through the nozzle filter at 100 psi. As the oil passes through the slots in the distributor into the swirl chamber, the pressure energy is converted into velocity energy that sends the oil spinning toward the orifice. Centrifugal force exerted on the oil creates an air opening in the center of the oil as it enters the orifice. The result is a hollow

Figure 6-17 **Cutaway view of oil burner nozzle.** *Courtesy: Delavan.*

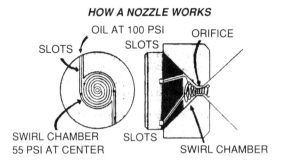

Figure 6-18 **This diagram shows how the oil pressure is converted to velocity inside the nozzle swirl chamber.** *Courtesy: Delavan.*

tube of oil that becomes a cone as it leaves the orifice where the film of oil explodes into tiny droplets. Figure 6-18 illustrates the conversion of pressure energy into velocity energy and the resulting action in the swirl chamber.

NOZZLE SELECTION—REPLACEMENT

It is essential to install the correct nozzle in every oil burner. It is equally essential that the service technician does not arbitrarily change the size, angle, or spray pattern of the nozzle when he has to replace it. The installation of the wrong nozzle can reduce the operating efficiency of the burner, create soot and smoke, and lead to this complaint from the home owner: "I am using more oil this year than last and it has been warmer. What did you do to my burner?"

There are several methods of determining the correct nozzle size when replacing the nozzle on any installation. The first and most obvious, is to make an exact replacement. If you are there to do an annual tune-up and the job has been operating properly for some time, you can assume that the nozzle that is being replaced was the proper size for this job. If you are not sure that the nozzle being replaced was the right size you will have to determine the size by one of the following methods:

1. Check the boiler or furnace for a plate that may indicate the nozzle size or a Btu input for the boiler. If the plate indicates Btu input, the nozzle size can be calculated by the following formula:

$$\frac{\text{Btu Input}}{140{,}000 \text{ Btus}} = \text{Nozzle Size in GPH}$$

The 140,000 represents the amount of heat released by one gallon of fuel oil when it burns.

2. If the plate indicates Btu output for the boiler, use this formula:

$$\frac{\text{Btu Output}}{112,000 \text{ Btus}} = \text{Nozzle Size in GPH}$$

The 112,000 represents the amount of heat absorbed by a boiler operating at 80% efficiency. Experience will teach you that some boilers do not operate at 80% efficiency and you will have to adjust to the next larger size nozzle.

3. If the boiler plate indicates square feet of steam, you can convert this to Btus by multiplying by 240. Determine the nozzle size by using formula #2.

4. The square feet of steam rating can be converted to the approximate nozzle size by dividing by 300.

5. If the boiler plate indicates square feet of hot water at 180°F, convert to Btus by multiplying by 165, then follow formula #2.

6. If there is no plate on the boiler, measure the combustion chamber. Multiply the length by the width to determine the square inches and then divide by 90. This method will give you a nozzle size that will produce a flame that fits into the combustion chamber. Check it against the size of the nozzle that you are replacing. They should be very close to the same size.

Once the size is determined, you have to select the spray angle that will determine the shape of the flame. The flame should fit inside the combustion chamber without making contact with the side walls or the floor. If the flame hits either of these parts of the chamber, we will get a hard carbon buildup that will cause smoke and soot formation. Some of the flame retention head burners require a specific angle of spray. Check the burner label or installation instructions to find out if the burner does indeed have this requirement.

The fuel oil leaves the nozzle in a cone-shaped spray. There are three different spray patterns available in every nozzle size (see Figure 6-14): the solid, which has tiny droplets of oil throughout the entire cone; the semi-solid, which leaves a small space inside the cone where there are no oil droplets; and the hollow, which leaves a large space inside the cone where there are no oil droplets. The reason for these different spray patterns is that the fuel spray is matched to the air delivery pattern of the oil burner. Conventional oil burners, which have a stabilizer and end cone combination, deliver air in a solid pattern from the end of the burner. By using a solid nozzle with this type of burner, we are placing the droplets of oil directly into the air stream so that each drop can be surrounded by air and have the oxygen it needs to burn completely (see Figure 6- 19). The semi-solid nozzles are used with the flame retention burners for exactly the same reason as are the hollow nozzles with the shell head burners (see Figures 6-20 and 6-21).

Remember that you must be careful when replacing nozzles. If you install a nozzle that is too small, you will not heat the building properly and the burner will run continuously. If you install a nozzle that is too large, you may create a problem of impingement, in which the flame hits the combustion chamber, which will result in smoke and

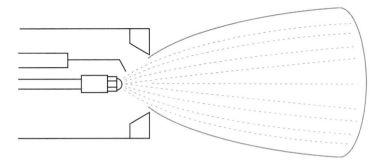

Figure 6-19 The solid nozzle spray pattern matches the solid air delivery pattern of a standard oil burner.

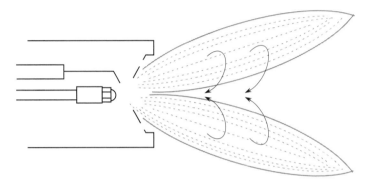

Figure 6-20 The semi-solid nozzle spray pattern matches the air delivery pattern of flame retention head oil burners.

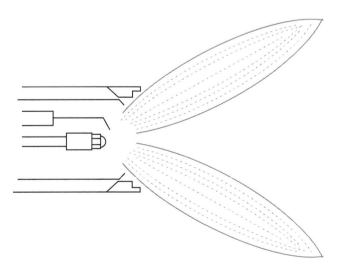

Figure 6-21 The hollow nozzle spray pattern matches the air delivery pattern of Shell head oil burners.

soot. In either case you will have an irate customer on your hands. The wrong spray angle can also lead to impingement. The wrong spray pattern can lead to your adjusting the burner for excess air, which will reduce the operating efficiency of the unit and get the customer upset. Too many homeowners have converted from oil to alternative fuels because of service technicians who either did not know or did not care enough to install the proper nozzle. There is one certain method of determining the exact nozzle for a particular installation and that is by performing the nozzle application test. This procedure requires time and experience. We will discuss nozzle application testing in Chapter 20, Improving Combustion Efficiency.

The nozzle must be treated with respect because of its importance in the operation of the burner. Always keep the nozzles in their plastic containers to keep them from becoming contaminated. Most service technicians have a nozzle box (see Figure 6-22) as part of their equipment so that they can keep their nozzle supply clean and easily accessible. It should be noted that all nozzles are tested before they are sent to the supply houses. The test will leave some oil in the nozzle when it is packed in its case. It is possible, particularly in cold weather, for this residual oil to congeal in the nozzle, which will result in clogging the nozzle. If you are going to replace a nozzle that has been outside in your service vehicle, put the nozzle in its protective case in your pocket for a while to warm it up. This can save you a considerable amount of time and avoid returning perfectly good nozzles to the supplier because they are "defective."

Figure 6-22 **Delavan nozzle box protects the service van's inventory of nozzles and has space for vacuum and pressure gauges.** *Courtesy: Delavan.*

COLD OIL AND NOZZLE PERFORMANCE

Cold oil from outside storage tanks can create problems with nozzle performance. As the temperature of the oil drops, the viscosity increases (see Table 6-2). Increasing the viscosity of the oil will slow down the velocity in the swirl chamber and create a thicker cone of oil as it leaves the orifice. This thicker cone of oil will actually increase the flow rate of the nozzle as well as creating much larger droplets of oil, as shown in Figure 6-23. The larger droplets of oil may be hard to ignite or may produce a long narrow flame. Increasing the flow rate of the nozzle without adjusting the air will result in smoke and soot formation.

	TEMPERATURE °F			
	100	**80**	**30**	**20**
Viscosity SSU	35	37	52	65

Table 6-2   Effects of Temperature on #2 Oil Viscosity.

Figure 6-23   **Reducing the pump pressure will result in larger droplets of oil leaving the nozzle. Cold oil will have a similar effect on droplet size.** *Courtesy: Delavan.*

The solutions to this problem are to preheat the oil or to increase the pump pressure. If the pump pressure is increased, the nozzle size should be reduced to maintain the correct firing rate.

NOZZLE PARTS

All the information about the nozzle is stamped on the side of the nozzle outer nut. The information includes:

1. The size, in GPH, at 100 psi.
2. The spray angle.
3. The spray pattern.

Example: 1.65   1.65 GPH
              60   60 degree spray angle
            S-S   Semi-solid delivery pattern

The parts of the nozzle are shown in Figure 2-24. The outer nut of the nozzle fits into a 5/8″ box wrench. The face of the outer nut is a polished surface where the orifice, a

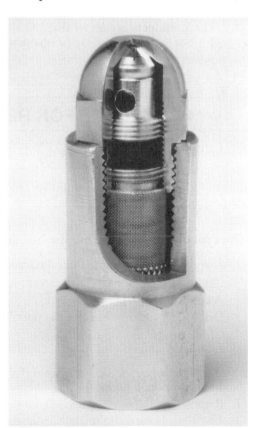

Figure 6-24 **Cutaway view of oil burner nozzle in nozzle adaptor shows the parts of the nozzle.** *Courtesy: Monarch Manufacturing Works, Inc.*

	DELAVAN	HAGO	MONARCH	STEINEN	DANFOSS
Solid	B	ES/P	R/AR	S	AS
Semi-solid	S-S	S-S	PLP	S-S	AB
Hollow	A	H	NS/PL	H	AH

Table 6-3   **Nozzle Interchange Chart.**

precision drilled hole from which the oil sprays, is located. This is actually the metering device, because each orifice will allow a specific amount of oil to pass through it at 100 psi. Directly behind the orifice is the swirl chamber. This is a small beveled cut in the nozzle into which the distributor of the nozzle fits.

The distributor is at the end of the screw pin or stem that screws into the nozzle outer nut. The distributor has slots that feed the oil into the swirl chamber at an angle so the oil will come out of the orifice in a spinning motion. The slots in the distributor are matched to the orifice so nozzle parts cannot be switched from one nozzle to another. Once the stem is installed in the outer nut, just slightly more than hand tight, the nozzle strainer is installed.

These strainers are either fine metal mesh or bronze sintered filters. They are designed to catch small particles of dirt that escape the pump strainer to keep the nozzle from getting clogged.

PROCEDURE FOR REPLACING THE NOZZLE

It is highly recommended that the nozzle be replaced at least once a year, usually during the annual tune-up. It is very difficult to clean a nozzle on the job. Even a small scratch or nick in the orifice will lead to improper atomization of the oil. The best solution to nozzle problems is a new nozzle!

The Procedure for Replacing the Nozzle Is as Follows:

1. Disconnect the power to the oil burner.
2. Remove the drawer assembly cover plate or open the ignition transformer, whichever is applicable.
3. Disconnect the oil line from the pump. Use the correct size open-end wrench.
4. Check the position of the drawer assembly in the burner.
5. Disengage any locking device—lock nut or screw—that holds the drawer assembly in position.
6. Disengage the high tension wires from the ignition transformer.
7. Remove the drawer assembly from the burner. Be careful not to spill the oil in the 1/8″ pipe. Have a can ready to pour the oil into.

8. Remove the nozzle from the drawer assembly using a nozzle extractor (see Figure 6-25) or two box-open end combination wrenches—a 3/4″ to hold the nozzle adaptor and a 5/8″ for the nozzle.

9. Rinse the nozzle adaptor and 1/8″ pipe with running water or connect the nozzle line to the fuel unit and pump some clean oil through it into a can. Install the replacement nozzle with the nozzle extractor or two wrenches.

10. Clean and check the electrodes and ignition system parts before installing the drawer assembly in the burner.

11. Reconnect the high tension leads to the ignition transformer and reconnect the oil line. Be certain that the locking device, which holds the drawer assembly in place, is tight and that the drawer assembly is back where it belongs.

12. Close the drawer assembly access opening.

13. Turn the power on and check for smooth light off, flame size, smoke, and oil leaks.

14. If the nozzle strainer was dirty, now is a good time to clean the pump strainer.

15. This is a good time to run an efficiency test on the unit. Efficiency test procedures are discussed in Chapter 18.

Figure 6-25 **The nozzle wrench has an extension for hard-to-reach nozzles. The drawer assembly must be removed from the burner before the nozzle can be extracted.** *Courtesy: Delavan.*

Figure 6-26 **Nozzle adaptors for various oil burners.** *Courtesy: Delavan.*

THE NOZZLE ADAPTOR

The nozzle is installed in the nozzle adaptor, which has a 1/8″ female pipe thread on one end and a 9/16″-24 universal thread on the other end (see Figure 6-26). There is an opening inside the adaptor for the oil to flow into the nozzle. On some adaptors this opening is near the top, which will prevent oil dripping through the nozzle after the burner shuts off. The adaptor is attached to a pipe that fits inside the burner. The other end of the pipe is connected to the fuel unit nozzle port with a piece of copper tubing. The stabilizer and the electrode bracket are on the same piece of 1/8″ pipe. When the nozzle is removed to be serviced, you can also service the ignition system parts.

DELAYED OPENING OIL VALVE

An optional part found on some burners is the delayed opening oil valve (see Figure 6-27). These valves are installed at the fuel unit nozzle port. This is an electrically operated valve that has a built-in delay mechanism. When power is applied to the valve, a solid state electronic device prevents the power from reaching the coil for approximately three seconds. Once the power reaches the coil, the valve opens and the oil can continue on to the nozzle. The three-second delay allows the oil burner motor to come up to full speed before the oil sprays out of the nozzle. This leads to a smooth light off of the flame. Another advantage is the positive closing action of the valve, which will ensure clean cutoff of the flame and eliminate the problem of oil dripping through the nozzle when the burner is shut off.

Figure 6-27 **Delayed opening oil valve—Honeywell V4046.** *Courtesy: Honeywell.*

OIL LINE CONNECTIONS

The oil line connection between the fuel pump and the nozzle assembly is usually made of copper or aluminum tubing. The connectors should be of the flared type rather than compression fittings (see Figure 6-28). Every service technician should be able to make a flared end on tubing because this is the most dependable type of connection that can be made when working with flexible oil lines. Do not use a pipe wrench or pliers to tighten the flare nut as these tools will strip the soft metal from which the nuts are made. Only the

Figure 6-28 **The flared copper tubing connection is recommended for all oil line connections.**

appropriate open-end wrench should be used to tighten these connectors. Be sure that the wrench is properly seated on the nut; push it all the way on to the fitting before you tighten or loosen the nut. Always check for leaks before you leave a job that you have taken apart. It only takes a second and can prevent a repeat service call. Wipe all tubing connections dry and wait a few seconds, then touch them with your finger to see if there is a leak. Clean up any oil that has spilled and take all oily rags or newspapers out of the building with you and deposit them in a garbage receptacle.

To Briefly Summarize

1. The nozzle is a precision instrument and must be handled carefully.
2. The functions of the nozzle are to:
   A. Atomize the oil.
   B. Meter the oil entering the combustion process.
   C. Deliver the oil in a specific spray pattern.
   D. Deliver the oil at a specific spray angle.
3. The nozzle spray delivery pattern must match the burner's air delivery pattern so that the burner can be set to operate efficiently.
4. Annual nozzle replacement is recommended.

Please Answer These Questions

1. How is the pressure test performed on a pump and why is it important?
2. How is the vacuum test performed on a pump and why is it important?
3. How is the cutoff test performed on a pump and why is it important?
4. When replacing a nozzle, why is it important to install the correct spray pattern for that particular burner?
5. When replacing a nozzle, why must you be careful to install the correct size for that particular installation?
6. What do these markings on the outer nut of a nozzle indicate? 2.50 60 S-S
7. Why is it important to clean a pump strainer before running tests on the pump?
8. What is the difference between a single-stage and a two-stage pump?
9. What must be done to a pump to set it for a two-pipe installation?
10. Why do the pump manufacturers suggest that Teflon tape not be used on oil line piping?
11. Why is it important to always replace the pump strainer cover gasket every time the strainer cover is removed?
12. Explain the procedure for replacing an oil burner nozzle.
13. What effect does cold oil have on nozzle operation?

CHAPTER 7

THE IGNITION SYSTEM

High pressure gun type oil burners depend on a high voltage electric spark to supply the heat required to start the combustion process. This is actually a chain reaction. As the air–oil mixture leaves the burner, the spark vaporizes and ignites a very small portion of the mixture. This initial flame is hot enough to ignite the rest of the mixture. When the system is operating properly, the light off is very smooth. However, a rough light off will cause vibration, smoke, odors, and a guaranteed upset home owner. Service technicians must be aware of the importance of the ignition system and take time to be certain that the system is operating properly every time the burner is serviced. Here are some of the things that can cause problems with the ignition system:

1. The setting of the electrode tips to deliver the spark where it can graze the air–oil mixture as it leaves the burner is absolutely essential to achieve smooth light off (see Figure 7-1). The oil spray must not touch the tips of the electrodes. If that happens, a carbon bridge will form between the tips and short-circuit the spark. The result will be a safety lockout of the burner. There are several diagrams of electrode settings supplied by burner manufacturers found in this chapter. It is always advisable to follow the instructions that are sent with the burner. Some manufacturers have devised electrodes that can fit in only the exact place to ensure smooth light off. This is an attempt to eliminate human error, which has been a serious problem with oil burners in the past. Remember, the spark will take the easiest path to a grounded surface rather than jump between the electrode tips. To prevent this, the setting of the distance between the tips must be smaller than the distance between any metal part of the electrode and a grounded metal part of the burner—like the flame retention head.

2. The air–oil mixture must be balanced to achieve smooth light off. If there is too much air or not enough oil, it will be difficult for the initial vaporization to take place. This will lead to a hard light off. Some things to check are:
   A. Is the air shutter open too wide?
   B. Is the retention head set properly?
   C. Is there too much draft, negative pressure, in the fire box of the boiler?

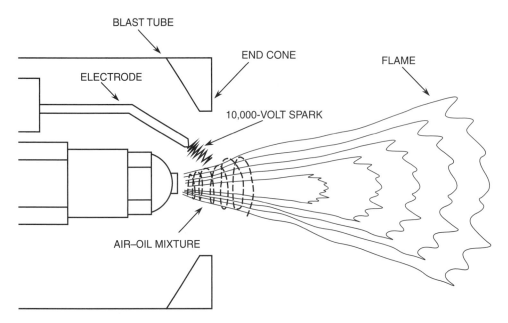

Figure 7-1 **The electrode tips are set so that the ignition spark grazes the edge of the air–oil spray as it leaves the end of the oil burner.**

    D. Is the entire fuel delivery system (nozzle, filters, pump pressure, etc.) working properly?

    E. Is the oil contaminated with water or is it too cold or heavy to vaporize?

 3. Are the parts of the ignition system operating properly? Detailed explanations of the operation of the system parts follows in this chapter.

    Understand that the smooth ignition of the air–oil mixture depends on all of the oil burner systems. This is where they all come together to create the flame that extracts the energy locked up in the fuel. From this point on you are dealing with fire. Be careful! Be extremely careful!

PUFF BACK

A *puff back* is an explosion caused by an accumulation of the air–oil mixture in the combustion chamber that is ignited by either a "late" spark or the hot refractory material. How serious the explosion is depends on the following:

 1. What is the firing rate of the oil burner? How much oil was in the combustion chamber when ignition finally took place?

 2. What is the safety timing of the flame safeguard control? How long did the burner run before ignition?

Regardless of the cause, the results are soot all over the house, boiler doors blown open, draft regulator blown out, flue pipe knocked down, or any combination of these. Puff backs can and must be avoided. Careful maintenance and service work are the best defense in preventing puff backs.

IGNITION SYSTEM PARTS

The Ignition System Has Only Four Parts:

1. The ignition transformer or electronic sparking device that provides the 10,000- to 14,000-volt spark.
2. The electrodes that deliver the spark at the exact point required for smooth light off.
3. High tension wires, bus bars, or springs that transport the high voltage electricity from the transformer to the electrodes.
4. The electrode bracket that keeps the electrodes in the proper position for smooth light off. On many older burners the stabilizer was used as the electrode bracket.

THE IGNITION TRANSFORMER

The ignition transformer is a step up transformer, which means that it increases the voltage from 120 to 10,000 volts. The operation of transformers is discussed in detail in Chapter 12, Electrical Equipment.

The Transformer Consists of (see Figure 7-2):

1. A heavy steel *can* that has a mounting plate, an electrical junction box for the primary coil connection, and an opening for the secondary coil connections.
2. A primary coil that has lead wires extending into the junction box.
3. Two secondary coils that have leads connected to the high voltage terminals.
4. A laminated, soft steel core, around which the coils are wrapped. The core provides the magnetic field required to operate the transformer.
5. Coils and core that are inside the can, which is then filled with a tar-like compound for insulation.
6. High voltage terminals surrounded by heavy, glazed porcelain insulators.

THE ELECTRONIC IGNITOR

Electronic ignitor spark generators (see Figure 7-3) supply a 14,000-volt spark to ignite the oil. The increased voltage also increases the temperature, which allows the spark to vaporize and ignite more of the oil in the air–oil mixture as it comes out of the blast tube.

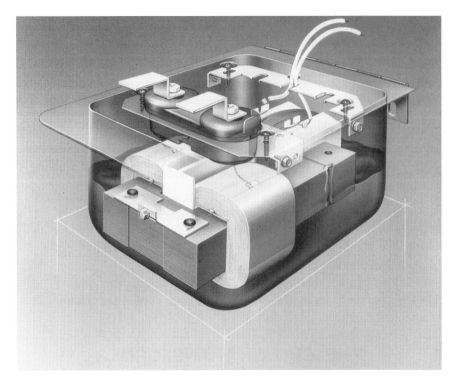

Figure 7-2 **Internal view of an ignition transformer.** *Courtesy: Webster Heating Products, Inc.*

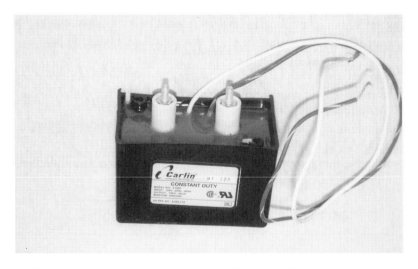

Figure 7-3 **The electronic spark generator supplies a 14,000-volt spark for ignition of the air–oil mixture.** *Courtesy: Carlin Combustion Technology, Inc.*

This results in a smooth light off even if the oil is cold or contains more of the heavy fractions resulting from the refining process. These units are well sealed are not susceptible to water damage the way transformers are. Another advantage over transformers is that they deliver the same 14,000 volts over a wide range of input voltages, 102 volts to 132 volts. You can be certain that if the motor can run, there will be a spark to ignite the oil.

Electronic Ignitors Are Solid State Units That Operate as Follows:

1. The 120-volt AC current is internally converted to DC.
2. The DC voltage turns power transistors on and off very quickly, sending current through the primary coil of a small internal transformer at a frequency of 15,000 to 30,000 Hz.
3. The secondary coil of the special high frequency transformer produces the high voltage ignitor output that also has a frequency of 15,000 to 30,000 Hz.
4. The high frequency allows the voltage to peak at the output voltage of 14,000 volts. This high voltage current supplies an intense, active spark at the electrode tips.

An electronic ignitor can be used as a direct replacement for an ignition transformer. It is attached to a mounting plate that fits on the oil burner in the exact position as the transformer (see Figure 7-4). There are ignitor kits available (see Figure 7-5) that include the spark generators and an assortment of base plates so the service mechanic can adapt an electronic unit to replace a transformer on any oil burner.

Figure 7-4 **Electronic ignitors on mounting plates for various oil burners.** *Courtesy: Carlin Combustion Technology, Inc.*

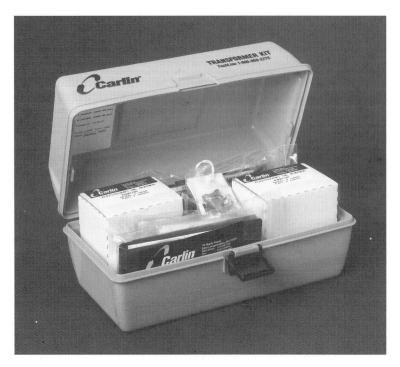

Figure 7-5 **Electronic ignitor replacement kit for service van inventory.** *Courtesy: Carlin Combustion Technology, Inc.*

The transformer mounting location will vary with the burner manufacturer's design. Figure 7-6 shows a universal replacement transformer kit that can be adapted to a variety of oil burners. Many new oil burners have the transformer mounted on top of the burner on a hinged mounting plate (see Figure 7-7). Access to the drawer assembly is achieved by unbolting the end of the transformer and swinging it open. On this type of burner, the secondary or high voltage terminals are readily available for inspection and testing. The spark generated by the transformer or sparking device must be transported to the electrodes. This is accomplished by using bus bars, springs, or high tension wire.

HIGH TENSION WIRE AND BUS BARS

Oil burners that use bus bars or springs are designed so that when the transformer is in the closed position the secondary terminals (high voltage) are in contact with the bus bars or springs. Bus bars are either metal rods or flat strips that are attached firmly to the end of the electrode (see Figure 7-8). The bars extend toward the rear of the burner where they come into contact with the secondary terminals of the transformer. Bus bars are not insulated, and special attention must be paid to not allowing them to get too close to any metal part of the oil burner. Electricity will always take the easiest path to ground. If the

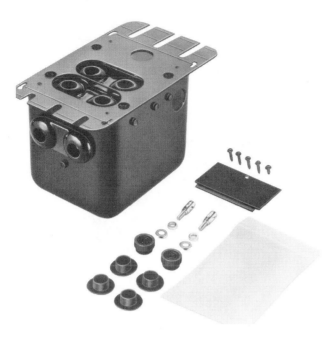

Figure 7-6 **Universal replacement ignition transformer can be adapted to fit a variety of oil burners and reduces the stock inventory of the service vehicle.** *Courtesy: Webster Heating Products, Inc.*

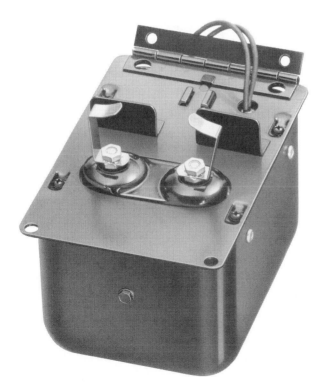

Figure 7-7 **Ignition transformer with hinged mounting plate for the Beckett AF oil burner.** *Courtesy: Webster Heating Products, Inc.*

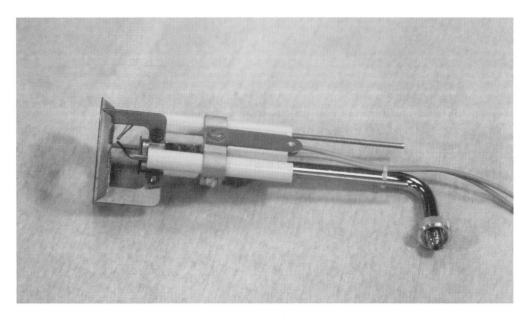

Figure 7-8   **The Carlin EZ-1 oil burner drawer assembly shows the positioning of the electrodes and bus bars.** *Courtesy: Carlin Combustion Technology, Inc.*

spacing between the electrode tips is larger than the space between the bus bar and a part of the burner, the spark will be inside the burner and not out at the electrode tips. Springs used for this purpose resemble the old screen door closing springs. They are not insulated and must not come into contact with each other or any parts of the burner. An additional problem with springs is that if they are too tight, they may pull the electrodes back into the burner. This will result in either delayed or nonignition and can be avoided by merely stretching the spring a little.

High tension ignition wire is #14 stranded copper wire covered with heavy plastic insulation. A variety of terminal ends are available for use in connecting high tension wire to both the transformer and the electrodes. The best of these is the type that has a metal sleeve that slips onto the outside of the high tension wire.

High Tension Wires Should Be Assembled as Follows (see Figure 7-9):

1. Strip approximately 3/4″ of the insulation off the end of the high tension wire.
2. Separate the strands of the exposed wire and bend them back along the insulation. Use a star pattern so there are strands of wire all around the exterior insulation.
3. Slide the terminal end sleeve over the strands of wire. Push the terminal all the way onto the wire and crimp the sleeve to keep it from coming loose.

This method ensures a solid connection between the high tension wire and both the transformer and the electrodes.

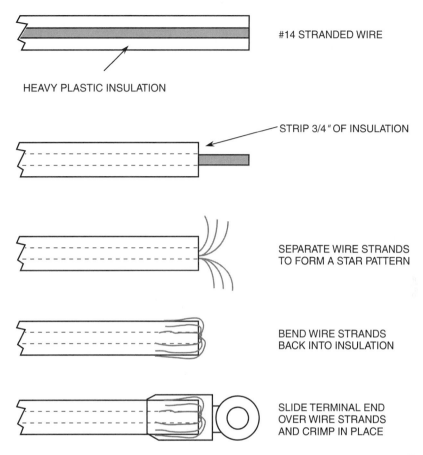

#14 STRANDED WIRE

HEAVY PLASTIC INSULATION

STRIP 3/4" OF INSULATION

SEPARATE WIRE STRANDS
TO FORM A STAR PATTERN

BEND WIRE STRANDS
BACK INTO INSULATION

SLIDE TERMINAL END
OVER WIRE STRANDS
AND CRIMP IN PLACE

Figure 7-9   **Procedure for attaching the terminal end to high tension wire for the ignition system.**

ELECTRODES

The electrodes are the oil burner's spark plugs. They are made of heavy steel wire that fits securely into a porcelain insulator (see Figure 7-10). Individual burner manufacturers design their own electrode so there are a variety of these in use. Regardless of their size or shape they all perform the same function. They deliver the spark to the exact position for smooth light off of the air–oil mixture as it leaves the blast tube. There are two types of porcelain, a ceramic material that is an excellent insulator used for electrodes, the glazed and the unglazed. Glazed porcelain has a shiny surface that is easy to clean while the unglazed is dull and hard to clean. The carbon that collects on the surface of the electrode porcelain is a conductor of electricity. It is possible for the high voltage electricity to track across the surface of the electrode and short out if there is enough carbon present.

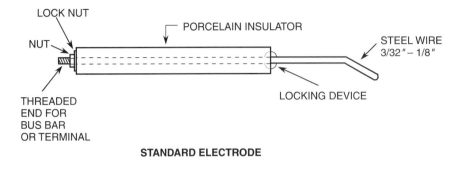

Figure 7-10 **Typical oil burner electrode construction.**

When servicing the ignition system, if you cannot get the porcelain white, change the electrodes!

The positioning of the electrodes is crucial to the proper operation of any oil burner (see Figure 7-11). Whenever possible, the burner manufacturer's instructions should be followed (see Figures 7-12, 7-13, and 7-14). If they are not available these general rules can be followed.

Proper Positioning of Electrodes:

1. The tips must be at least 1/2″ above the center of the nozzle. We do not want the oil spray to hit the tips. This will cause carbon to build up on the tips, which will affect the size of the spark. If the carbon buildup continues, it will actually bridge the gap and short-circuit the ignition system.
2. The tips should always be in front of the nozzle. How far in front depends on the angle of the nozzle spray pattern. Hold the drawer assembly up in front of your eyes and try to visualize the spray of oil coming out of the nozzle. The tips should be 1/8–1/4″ behind the spray to ensure no carbon buildup on the tips. The air coming through the blast tube will blow the spark into the edge of the spray and no oil will come into contact with the electrode tips.
3. The tips must be 1/8–3/16″ apart to get a spark large enough to ignite the air–oil mixture.

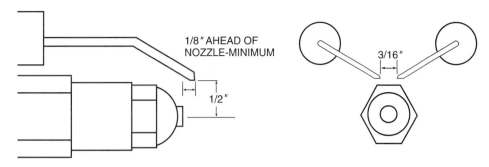

Figure 7-11 Standard oil burner electrode setting.

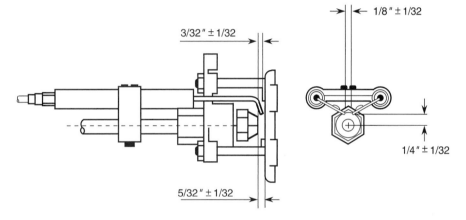

Figure 7-12 Electrode setting for the Beckett AF II oil burner. *Courtesy: R. W. Beckett Corp.*

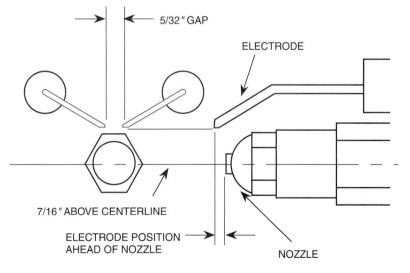

Figure 7-13 Electrode setting for the Beckett AF oil burner. *Courtesy: R. W. Beckett Corp.*

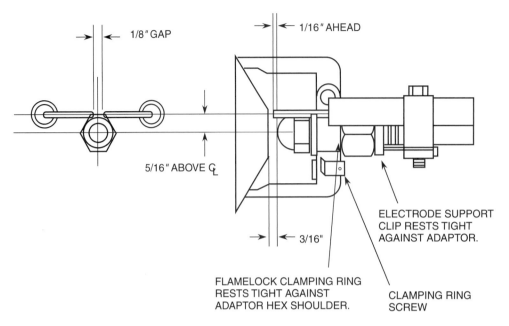

1/8" GAP   1/16" AHEAD

5/16" ABOVE C_L

3/16"

ELECTRODE SUPPORT
CLIP RESTS TIGHT
AGAINST ADAPTOR.

FLAMELOCK CLAMPING RING
RESTS TIGHT AGAINST
ADAPTOR HEX SHOULDER.

CLAMPING RING
SCREW

Figure 7-14   **Electrode setting for the Wayne Blue Angel oil burner.** *Courtesy: Wayne Home Equipment.*

4. If the burner is equipped with a flame retention head, the tips must be farther from the head than the spacing between them.
5. Check the air shutter adjustment to see how much air is going through the burner. A large quantity of air will blow the air–oil mixture away from the spark which means that the tips will have to be set farther in front of the nozzle.

SERVICING THE IGNITION SYSTEM

The ignition system must be serviced during the annual tune-up or any time that the burner locks out on safety. A safety lockout indicates that the burner attempted to start but there was no flame. At this point, you cannot assume that the ignition is functioning properly and that something else caused the lockout. The drawer assembly must be removed and the entire ignition system, electrodes, bus bars or high tension wire, and the transformer must be checked.

To Test the Ignition Transformer:

1. Disconnect the power to the oil burner.
2. Open the transformer or take it off the burner chassis so that you can see the high voltage terminals.

3. Clean the porcelain insulators around the high voltage terminals. Check for signs of cracking or crazing, which is a series of fine cracks in the porcelain.

4. Disconnect the power to the motor so that no oil will spray into the combustion chamber while you are checking the transformer.

5. Turn the burner power on and energize the primary control. Be careful. The transformer is alive and can give you a nasty sting if you touch one of the high voltage terminals!

6. Check the power at the transformer primary leads in the junction box. Use a volt meter. There must be 120 volts, or close to it, going to the primary to get the desired high voltage from the secondary. This must be done quickly because the primary control will lock out on safety, hopefully, in 15–30 seconds. If there is a safety lockout, wait a few minutes before resetting the control so that you can continue the test.

7. Test the secondary voltage with a high voltage tester (see Figure 7-15). Attach the alligator clip of the tester to the transformer case. Turn on the power to the transformer and touch the high voltage terminal with the tester probe. The neon bulb on the tester will glow. Turn the tester dial until the bulb goes out. Read the high voltage output of the transformer by multiplying the dial setting by 100. Each high voltage terminal should deliver 5,000 volts. This can also be done by touching a well-insulated screwdriver blade to the two terminals and gradually increasing the gap between the screwdriver and one of the terminals while still in contact with the other terminal. A good transformer can produce an active spark 3/4″ long. Be sure to use a dry screwdriver that has a solid plastic handle (see Figure 7-16). Do not use a screwdriver that has a shank that extends through the handle and sticks out of the end.

8. Do not hesitate to replace the transformer if you feel that it is weak. Do not hesitate to replace a transformer that has been under water. They cannot be dried out and used again.

Figure 7-15 Using a high voltage tester to check an ignition transformer. Each terminal should read 5,000 volts.

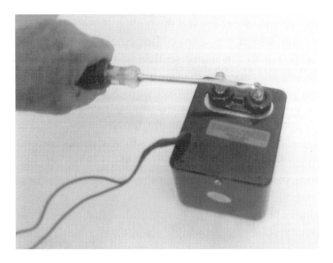

Figure 7-16 **Using a well-insulated screwdriver to test an ignition transformer. There should be an active spark across the terminals at least 3/4″ long if the transformer is good. Be extremely careful not to touch the terminals and use a screwdriver with a heavy plastic handle!**

9. If the transformer is good, reconnect the wiring to the motor, reattach or close the transformer, and run the burner.

To Test the Electronic Spark Generator:

1. Disconnect the power to the oil burner.
2. Take the spark generator off the burner chassis so you can see the high voltage terminals.
3. Clean the porcelain insulators around the high voltage terminals. Check for signs of cracking or crazing, which is a series of fine cracks in the porcelain.
4. Disconnect the power to the motor so that no oil will spray into the combustion chamber while you are checking the spark generator.
5. Turn the power on and energize the primary control. You may have to disconnect one of the cad cell leads to get the primary control to start. Be careful The spark generator is alive with the power on and can give you a nasty sting if you touch one of the high voltage terminals.
5. Using a well-insulated, dry screwdriver blade, touch one of the high voltage terminals and then extend the blade across to the other high voltage terminal. A strong, active spark should appear as you get close to the second terminal. Stretch the spark by moving the screwdriver away from the second terminal. The spark has to be at least 3/4″ long.
7. Another method of testing an electronic spark generator is with an Ohmmeter. Disconnect the wiring to the spark generator and test each high voltage terminal to the

burner chassis. You should get a reading of less than 2,000 ohms. The other terminal should give you the same, or close to the same, reading. The terminal-to-terminal resistance should be twice the individual terminal resistance. If the terminal to chassis resistances differ by more than 20%, the spark generator should be replaced.

Any time there is a safety lockout, or during the annual tune-up, the electrodes must be checked.

To Service the Electrodes Follow These Steps:

1. Disconnect the power to the oil burner.
2. Remove the drawer assembly.
3. Check the electrode tips for signs of carbon buildup, spacing, and their position in relation to the nozzle.
4. Remove the electrodes from their bracket. This must be done to ensure that the porcelain is not cracked inside the bracket. Clean the porcelain. Fuel oil can be used as a solvent to remove carbon from the surface of the porcelain.
5. Check for cracks in the porcelain. They will appear to be black lines that cannot be cleaned off.
6. Check for crazing of the porcelain. Crazing is a series of fine lines that resemble fingerprints on the surface of the porcelain. Cracks or crazing will allow the high voltage electricity to leak to ground and will short-circuit the ignition system.
7. Lock the electrodes back into their bracket. Set the tips as explained above or as required by the burner manufacturer (see Figure 7-17). Some electrode brackets have

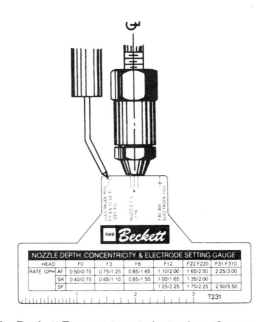

Figure 7-17 **Using the Beckett T-gauge to set electrodes.** *Courtesy: R. W. Beckett Corp.*

a split clamp with a screw that pulls the clamp tight around the porcelain. Do not over-tighten this screw because you don't want to strip the bracket. You would then have to contend with a nut and bolt to hold the electrode in place. Some electrode brackets have a set screw that directly holds the porcelain. Most of these require a metal bushing between the set screw and the porcelain. If the electrodes do not fit tight in the bracket, they may have to be shimmed. Aluminum pipe covering straps make an excellent shim stock for this purpose.

8. Clean the surface of the bus bars to ensure solid contact with the high voltage terminals of the transformer. Check the springs' tension and clean them. Check high tension wire for signs of insulation failure; check for loose terminals, then clean them.
9. Replace the nozzle while the drawer assembly is out of the burner.
10. Reassemble the burner. Pay particular attention to the position of the high tension leads from the transformer to the electrodes.
11. Run the burner and check for smooth light off. Use a flame mirror to check the setting of the electrode tips. You can see the oil spray graze the spark if the tips are set correctly. Run the burner through several starting cycles to test the ignition system. Check for oil leaks and clean up around the burner before you leave. Remove any waste from the building when you leave.

To Briefly Summarize

1. The ignition system supplies a high voltage spark that ignites the air–oil mixture as it leaves the end of the blast tube.
2. The setting of the electrode tips is crucial to the proper operation of the oil burner.
3. The burner manufacturer's instructions must be followed when setting the electrodes.
4. The entire ignition system must be checked during the annual tune-up and anytime the burner locks out on safety.

Please Answer These Questions

1. What is the secondary voltage of an ignition transformer?
2. Why does a weak transformer have to be replaced?
3. What is the material used for high voltage insulators?
4. What does the term *crazing* mean?
5. Why are cracks and crazing a serious ignition problem?
6. Explain how you would attach terminal ends to high tension wire.
7. What effect would oil hitting the electrode tips have on the ignition system?
8. What are the factors that determine the position of the electrode tips in relation to the nozzle to ensure smooth light off?
9. What is a *puff back*?
10. When should the ignition system be serviced?

CHAPTER 8

THE STEAM HEATING SYSTEM

Steam heating systems can be found in a variety of buildings, from the smallest residence to a giant skyscraper. There are several different types of steam systems such as the single-pipe, the two-pipe, the vapor vacuum, and the vacuum return systems. All of them operate on the basic principle that when water is heated to its boiling point, 212°F, or at a lower temperature in a vapor vacuum system, it changes from a liquid into a gas. As you heat the water, it will absorb the heat and continue to change into a gas until enough steam is created to fill the system. As the water is changed into steam, it expands to 1,650 times its original volume. This means that a small amount of water can make a lot of steam.

Steam has many uses besides heating buildings, such as generating electricity, supplying the power for ships, cleaning, industrial processes of all kinds, air conditioning, and production of hot water. Steam is a faithful but sometimes dangerous servant of mankind. Working with steam-producing equipment is an inseparable part of this industry. The following definitions will help you to follow the text:

1. **British Thermal Unit (Btu)**—The amount of heat required to raise the temperature of one pound of water one degree Fahrenheit.
2. **Boiling Point**—The temperature at which water begins to turn into steam. At sea level, the boiling point is 212°F. Reducing the pressure on the water, as in a vapor vacuum system, will lower the boiling point temperature. Increasing the pressure on the water will raise the boiling point temperature.
3. **Steam**—An odorless, colorless gas formed by heating water to a temperature of 212°F or higher.
4. **Pounds Per Square Inch (psi)**—The indication of the pressure inside the steam system.
5. **Latent Heat of Steam**—The amount of heat, in Btus, required to convert one pound of water at 212°F into steam at 212°F. The latent heat of steam is 970 Btus.
6. **Condensate**—Water that is formed when steam releases the latent heat and changes back into water.
7. **Water**—The water used in the production of steam must be free of all contamination. Only potable (drinking) water can be used. Salt, lime, or dirt in the water will raise the boiling point.

8. **Saturated Steam**—Steam that has just formed and will return to water quickly as it loses some heat.
9. **Superheated Steam**—Steam that has additional heat and will not return to water quickly.

You can produce some steam by putting water into a pot and heating the water on your gas or electric range. As the water absorbs the heat, you will see bubbles forming. As the water gets to its boiling point, 212°F, you will see the bubbles getting larger and moving to the top of the water. The movement of the bubbles will cause the water to surge in the pot. Hold the pot cover at an angle above the pot. You will see little drops of water forming inside the cover. The steam coming out of the water will condense on the cold metal of the cover. If you look carefully, you will see the steam condensing in the air. This looks like a cloud or fog forming above the pot. You cannot see the steam but you can feel it if you hold your hand above the pot for a second. Be careful not to burn yourself, and remove the pot from the flame before all the water evaporates. If you were to leave the pot on the flame with no water in it, you would melt the pot. This happens because there is no longer any water in the pot to absorb the heat from the flame.

Steam systems require a continuous supply of clean water. All steam systems are connected to the building water supply system. The heat energy used to turn the water into steam comes from the fuel. Any of the fuels can be used for this purpose, but we will be discussing #2 fuel oil because this is the fuel used for residential systems. The container that holds the water while it is being heated is the steam boiler.

STEAM BOILERS

There are two basic types of boilers used for domestic heating systems: the steel fire tube boilers and the cast-iron sectional boilers (see Figure 8-1). All boilers have these parts:

1. The fire box, where the fuel is burned.
2. The flue gas passages, where the hot flue gas releases heat for the water in the boiler to absorb.
3. The water jacket, the hollow portion of the boiler where the water is held.
4. The steam chest, a space at the top of the water jacket where steam can collect (see Figure 8-2).

STEAM BOILER TRIM

Steam systems can be identified by the steam boiler trim. This consists of the steam pressure gauge, the steam safety valve, and the gauge glass assembly. The steam trim package comes with the boiler when it is ordered from a supply house.

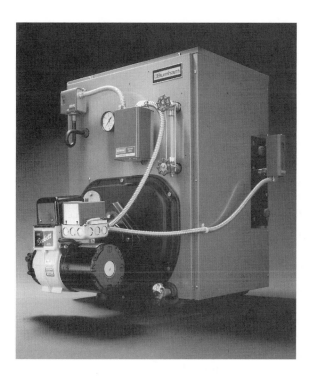

Figure 8-1 **Oil-fired steam boiler, Burnham V-7, shows location of boiler trim and controls.** *Courtesy: Burnham Corp.*

Figure 8-2 **Cut-away view of Burnham V-7 boiler shows the combustion chamber, flue gas passages, water jacket, and the location of the internal domestic hot water tankless coil.** *Courtesy: Burnham Corp.*

The Steam Boiler Trim Is Installed as Follows:

1. The steam pressure gauge is installed in a special tapping on the boiler that leads into the steam chest of the boiler. These gauges read from 0 to 15 psi. The gauge will show the pressure in the system during a heating cycle. Residential systems operate in the 2–5 psi range, depending on the building. The gauge must read 0 before any piping on the boiler or any part of the heating system can be disconnected.

2. The steam safety valve (pop-off valve) is installed in a special tapping on the boiler that leads into the steam chest of the boiler. These safety valves are set to open at 15 psi and are designed to release steam rapidly to prevent a boiler explosion. The valve must be the same size as the boiler tapping and cannot be reduced in size. Side outlet valves are recommended so that the discharge from the valve can be directed away from an area where someone might be injured (see Figure 8-3) or piped to discharge 6″ above the floor.

3. The gauge glass assembly consists of a Pyrex glass tube that fits into two gauge cocks. One gauge cock is installed in a special tapping in the steam chest of the boiler. The other gauge cock is installed in a special tapping that is below the water line of the boiler. These tappings are set so that when the glass shows half full you have the proper amount of water in the boiler. The glass tube is sealed by heavy rubber washers that fit into the nut on the gauge cock. The gauge cocks can be closed so that you can clean or replace the Pyrex tube without draining the water out of the boiler. The normal setting for the gauge cocks is wide open. This allows the glass to indicate the exact amount of water in the boiler (see Figure 8-4). Figure 8-5 shows the location of the trim items as well as the near boiler piping.

Figure 8-3 **Angle steam safety (pop-off) valve.**

Figure 8-4 **Boiler gauge glass assembly.**

THE SINGLE-PIPE STEAM SYSTEM

The single-pipe steam system requires carefully designed piping to operate properly. The system gets its name from the fact that steam and condensate are in the same pipe at the same time. The piping system has no extra parts and is installed in the order listed below (see Figure 8-7):

1. The steam header is made of pipe and fittings the same size as the steam outlet of the boiler. The bottom of the header must be at least 18″ above the water line of the boiler. When replacing an existing boiler, try to get as much height as possible, 24–36″, between the water line and the header. New boilers have small steam outlets; this leads to an increased velocity of steam leaving the boiler. The increase in velocity has a tendency to carry water with the steam. This creates a problem with some of the new cast-iron steam boilers that have a single steam outlet in the front section. Many of these boilers do not have a combustion chamber in the firebox. Because of this there is extremely rapid heating on the water, which results in high steam velocity leaving the boiler through the single outlet. This rapid movement causes the water in the rear section to drop and exposes portions of that section to a "dry fired" condition, This will result in a cracked boiler section. To prevent this, it is necessary to install an insulating material on the floor of the fire box, or a complete combustion chamber, to slow

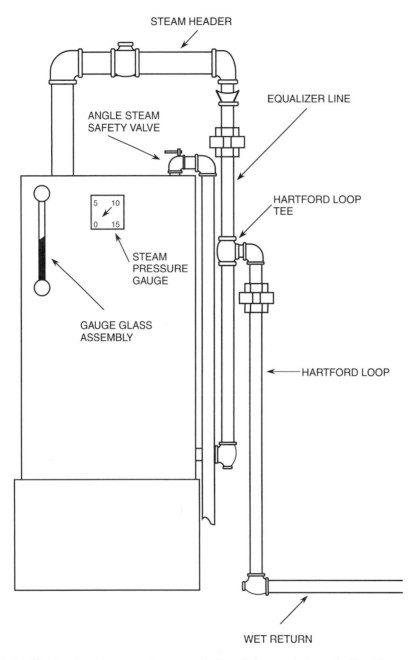

Figure 8-5 **Single-pipe steam system near boiler piping and steam boiler trim.**

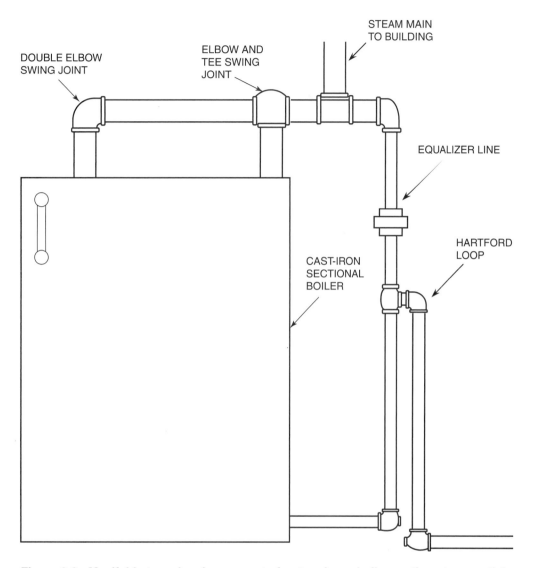

Figure 8-6 **Manifold steam header connects front and rear boiler section steam outlets. Swing joints must be used to avoid cracking the boiler.**

down the heat exchange between the burner flame and the water in the boiler. If the boiler has two steam outlets, one in the front section and the other in the rear section, they must be connected with a full size manifold (see Figure 8-6). Swing joints— combinations of fittings connected by short pipe nipples—must be used to provide flexibility in the piping. This is necessary because as the boiler produces steam, there is some expansion in the cast iron of the boiler sections and the steel piping. These materials do not expand at the same rate, and if there is no movement in the piping,

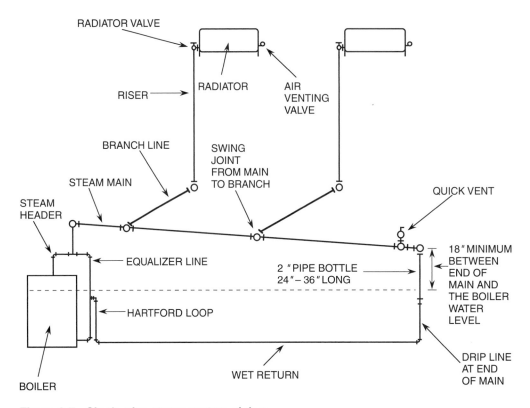

Figure 8-7   **Single-pipe steam system piping.**

the boiler will develop cracks around the steam outlets. The take-off for the steam mains must come at the end of the manifold and not in the middle of it. The basic purpose of the steam header is to collect dry, superheated steam before it moves into the steam mains.

2. The equalizer line starts at the end of the header and connects the header with the bottom of the boiler. This pipe has three functions.
   A. It equalizes the pressure from the top to the bottom of the boiler. This helps to keep the water level in the boiler steady and prevents water from the boiler backing up into the return piping.
   B. It acts as a drip line for the condensate formed in the header.
   C. It helps the condensate return to the boiler. The smallest size pipe that can be used for the equalizer line is 1-1/4″ pipe. The Hartford Loop tee is installed on this line. The Hartford Loop tee is designed to prevent all the water from leaking out of the boiler if their is a leak in the wet return piping. The center of the tee must be 2–4″ below the water line. The return line connection into the tee is made with a short nipple.

3. The steam mains start at the header and are run out around the basement. The purpose of the steam mains is to distribute the steam equally to all parts of the building. The

smallest size pipe that is used for mains is 2″. The piping is installed so that it pitches—slants down away from the boiler. In order to get the required pitch, 1/2″ for every 10′ of run, swing joints must be used. The swing joints also allow the piping to move when steam enters it and the pipe heats up and expands. The expansion rate for steel steam piping is 1/16″ for every 100′ of run. Steam mains must be supported so that they do not sag and create traps where water can collect. The best method of supporting piping is by using Clevis hangers suspended from threaded rods. Quick vents are installed at the end of each steam main to release air as the steam enters the piping. Tees are installed along the mains for the branch lines that lead to risers that allow the steam to flow up to the radiators. The end of the steam main should be at least 18″ above the water level of the boiler (see Figure 8-8).

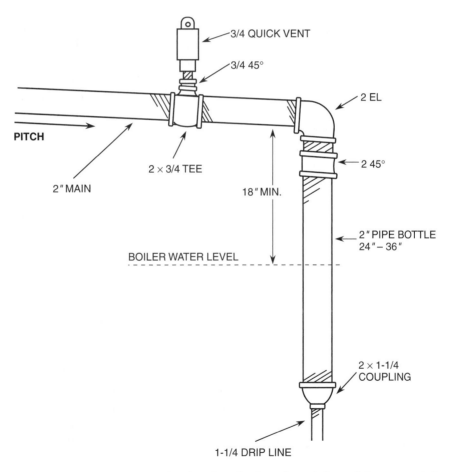

Figure 8-7 **End of the steam main showing the location of the quick vent and the drip bottle.**

4. Drip lines are pipes that connect the steam main to the return piping. The purpose of the drip lines is to drain condensate out of the steam main. On long mains there may be several drip lines. If the steam main has to be piped around an obstruction that will create a trap where water can collect, a drip line is installed to drain the trap. The drip line at the end of the main should start with a 2″ pipe bottle at least 3′ long. This will help to keep condensate out of the end of the steam main and will supply a little extra push to help the condensate return to the boiler. A 1-1/4″ pipe is usually used for drip lines.

5. Return piping can be wet, below the water line of the boiler, or dry, above the water line. Dry returns are usually quite short as they must pitch back to the boiler and if they are too long, they hang down too low and get in the way of any activity in the basement of the building. Wet returns can be run on the surface or buried in the ground. When return piping is buried, it must be covered with clean fill, not concrete or ashes that could eat through the pipe and cause a leak. The wet returns are connected to the Hartford Loop tee, which is on the equalizer line, 2–4″ below the water line of the boiler. The Hartford Loop is designed to prevent all the water from leaking out of the boiler if there is a leak in the wet return piping. There must be a drain valve at the end of every wet return pipe so that sediment can be cleaned out periodically.

6. Branch lines start at the steam main and run to the outside walls of the building. Swing joints must be used and the branch lines must pitch back to the steam main. Branch lines have to be supported so that they do not sag and cause traps where water can collect. Water trapped in any part of the system piping will cause a loud hammering noise that is extremely annoying.

7. Risers are pipes that run vertically through a building. Risers must be supported by riser clamps or couplings and floor plates. The radiators are connected to the risers. Very often the piping for this connection will be done in the space between the floor and the ceiling of the floor below. Swing joints must be used in this connection and special attention must be paid to the pitch as it easy to get an annoying trap under the floor. A riser can be used for several radiators on different floors. The size of the riser depends on the number and size of radiators connected to it.

8. Radiators for steam systems can be conventional tube-type cast iron or convector radiation. They are connected by using a steam radiator valve (see Figure 8-9). This type of valve has two parts: the valve and the spud. The valve has a large washer and seat and is either left completely open or closed tight. Trying to control the amount of heat at the radiator by partially closing the valve will only cause water to build up in the radiator, creating water hammer. The steam entering the radiator and the condensate draining out of it must pass through the valve at the same time. That is why the valve must be open as wide as possible. The spud is actually half of a union and requires a special tool to install it in the radiator (see Figure 8-10). The radiator must pitch back to the riser to allow the condensate to drain back toward the riser (see Figure 8-11).

9. Venting valves are used on each radiator and at the end of the steam mains to remove air from the system. These valves are open as long as there is air present in the radi-

Figure 8-9 **Angle steam radiator valve with spud.**

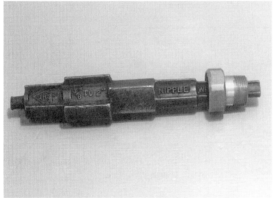

Figure 8-10 **A spud wrench must be used to tighten the spud in the radiator.**

ator or piping. They close as soon as steam enters the valve, which allows the system to build up pressure. It is possible to equalize the heat in the building by installing venting valves with orifices of varying sizes or adjustable venting valves. The radiators closest to the boiler would have a small opening while those farther away would have larger openings. The largest opening valves are the vents on the steam mains often referred to as *quick vents*.

SINGLE-PIPE STEAM SYSTEM OPERATION

The steam system operates automatically throughout the day in a series of cycles that are always the same. The thermostat calls for heat, which starts the oil burner. The heat released by the burning fuel is absorbed by the water in the boiler. As the water reaches the boiling point and more heat is released, the water begins to boil and turns into steam. Most

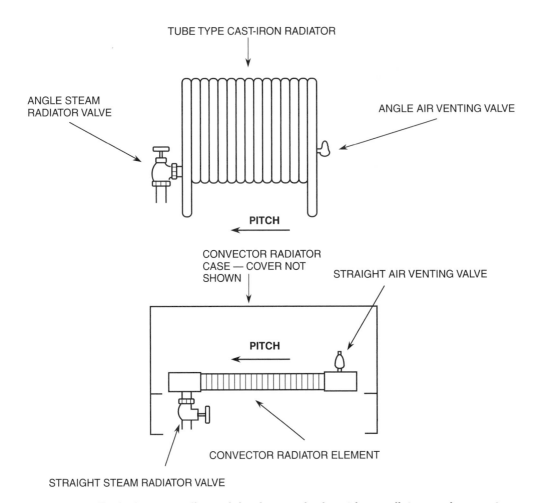

Figure 8-11   **Typical steam radiator piping for standard cast-iron radiators and convectors.**

boilers are built so a small percentage of the total water in the boiler has to be converted
to steam to fill the system. As the steam is formed in the boiler, it starts to expand and
moves into the steam header. Because this piping is cold, some of the steam condenses
and falls through the equalizer line back into the boiler. As heat continues to be released
into the boiler, the steam continues from the header into the steam mains. Whenever the
steam meets the cold piping, it will lose the latent heat and condense into water. The steam
and condensate will continue to move through the steam mains with the steam filling
most of the pipe and the condensate rolling along on the bottom of the pipe. As the piping heats up, there will be less and less condensate formed. The burner continues to burn
the fuel, releasing heat into the boiler, which causes the steam to continue filling up the
branch lines, risers, and eventually the radiators. Air that is trapped in the steam piping
is released through the quick vents on the steam mains and the air vents on each radiator.
Once all the air is released, the steam will fill the entire system. At this point, if the burner

continues to operate, steam pressure will start to build up in the system. Residential steam systems require very little pressure; 2–5 psi is usually enough to ensure that the entire system is full of steam. The hot radiators will now warm the air in the room by convection. Once the thermostat is satisfied, the burner will shut off, all the steam in the system will condense, and the condensate will return to the boiler so that the boiler will be ready for the next heating cycle. There are limit controls, low water cutoff and steam pressure control, operating controls, the thermostat, and possibly a hot water control wired in a circuit that allows this system to operate safely and automatically through many cycles a day. It is possible to generate domestic hot water with a steam heating system. The production of domestic hot water is covered in Chapter 17, Domestic Hot Water.

It is not the intention of this book to cover all of the variations of steam heating systems. For the most part, steam systems are obsolete, due to high operating cost and maintenance problems, and can be found only in older buildings. Steam is widely used for industrial purposes of all kinds. It is quite common for fuel oil dealers to have as accounts some form of industrial steam users. There may be a similarity in the burners and controls but you must be extremely careful when dealing with steam at the pressure required for an industrial purpose. Do yourself a big favor and do not attempt to repair or take apart something that you are not familiar with. Don't be bashful; ask for help if you need it.

To Briefly Summarize

1. Steam systems can be identified by the steam boiler trim, which consists of the pressure gauge, the gauge glass assembly, and the steam safety valve.
2. The single-pipe steam system is one of the systems found in residential buildings.
3. Steam piping must be installed to allow condensate to flow back to the boiler without traps that will cause water hammer.
4. Steam is used for industrial processes as well as for heating buildings. Many fuel dealers have industrial steam accounts.
5. Residential steam systems are obsolete and found only in older buildings.

Please Answer These Questions

1. What is the latent heat of steam?
2. Identify the four major areas in a steam boiler and explain their functions.
3. Explain the functions of the equalizer line.
4. What are swing joints and why are they used?
5. What are drip lines and where are they installed?
6. What are the three parts and functions of steam boiler trim?
7. Explain how a steam heating system can be balanced to provide even heat in a building.
8. What is the Hartford Loop and why is it important?
9. What does the term *pitch* mean?
10. What are *water hammers* and how can they be eliminated?

CHAPTER
9

HOT WATER
HEATING SYSTEMS

Hot water heating systems were developed to correct some of the problems of steam heating systems in residential applications. These were uneven heating, noise, keeping the water at the proper level in the boiler, and as the cost of fuel shot up, the expense. Steam systems require a change of state, from a liquid to a gas, during every heating cycle. The change requires an additional 970 Btus for every pound of water that must be changed from a liquid to a gas. This really adds up to a lot of extra fuel used during the heating season, fuel that can be saved by just heating water and circulating it through the radiation.

GRAVITY SUPPLY HOT WATER HEATING SYSTEM

The original hot water heating systems were based on the principal that the density of water decreases as it is heated, and the water becomes lighter. By installing a boiler in the basement and connecting supply and return piping between the boiler and the radiators, a system was designed to take advantage of this fact. Heated water would rise through the supply piping to the radiators. As heat is transferred from the radiators to the surrounding air, the water cools. The water becomes heavier and returns to the boiler through the return piping. The circulation of hot water in this manner, technically referred to as a thermosiphon, did effectively heat the building. This type of system, known as the gravity supply system, utilized coal as its fuel. The coal fire would be started early in the heating season and could be kept going throughout the winter. This resulted in a slow, even heat that was quite comfortable and easy to control.

The Parts of the Gravity Supply System Are as Follows:

1. **The Boiler**—The same boilers that are used for steam heating systems with a hot water trim package.
2. **The Hot Water Trim**—This consists of a thermometer, a pressure gauge, and a hot water relief valve that opens at 30 psi.
3. **Supply Main**—Large size pipe, 3″ or larger that is connected to the top of the boiler and runs through the basement to distribute the water to the radiators. The

large diameter pipe is used to reduce pipe friction, which would slow down the gravity circulation.

4. **Return Main**—The same size pipe as the supply main, it runs parallel to the supply main and is used to allow returning water from the radiators a method of entering the bottom of the boiler.

5. **Supply Risers**—The radiators are connected to the supply risers by means of a hot water radiator valve. The valves are installed at the top of one end of the radiator and used to control the flow of water to the radiator.

6. **Return Risers**—These are connected to the bottom of the radiator on the end opposite the supply connection.

7. **Radiators**—Large cast-iron radiators are used. These radiators have to be at least 30% larger than steam radiators for the same size room because this type of system does not get the radiators as hot, consequently a larger heating surface is required to heat the room.

8. **Air Bleed Valves**—These valves are installed on every radiator, usually on the return connection end. The purpose of these valves is to bleed all the air out of the system as it is being filled with water. This must be done because if one of the radiators is not full of water there will be no circulation through it and it will not get hot.

9. **The Expansion Tank**—This tank is located above the highest radiator, usually in a closet or in the attic of the house. An overflow line is connected to the top of the tank to allow excess water to drain out, usually onto the roof or into a leader pipe. This tank is connected to either a supply or return riser. The system is filled at the boiler by turning the water supply on. The water then flows through the boiler into the piping. As the system fills up, each radiator has to be bled to remove air trapped in the system. This takes a long time to do as there is a lot of water in these systems. Once the top floor radiators are full the water feed valve at the boiler is closed. There were some expansion tanks installed with a float-operated valve and water connection to keep the system full of water (see Figure 9-1).

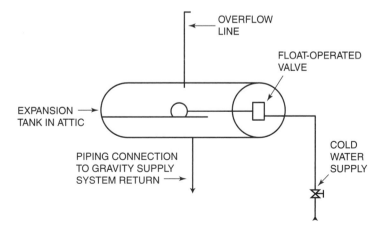

Figure 9-1   **The float-operated valve in the attic expansion tank keeps system full of water.**

Operation of the Gravity Supply Hot Water System Follows (see Figure 9-2):

1. A coal fire is started in the boiler (1).
2. As the water in the boiler is heated, it circulates by thermosiphon into the supply main (5), up the supply risers, and enters the radiators (9).
3. As hot water enters the top of the radiator, the colder water in the radiator moves through the bottom connection into the return riser (11), continues down into the return main (6), and then back to the boiler where it is heated and rises.
4. The amount of heat in each room can be controlled by adjusting the supply valve (8) on each radiator.
5. The amount of heat in the building can be controlled by adjusting the coal fire.
6. This control over a coal fire is not difficult. To cool the flame, you bank the fire by adding coal and closing the air supply. To increase the temperature of the water in the boiler, you must open the air supply to get the coal burning at a higher temperature.

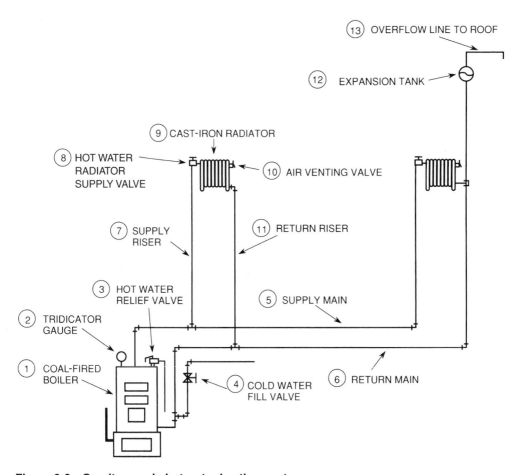

Figure 9-2   **Gravity supply hot water heating system.**

An experienced person can control this type of system and maintain a comfortable temperature in every room in the house.

Problems with this system are the extraordinary amount of piping in the basement, the slow pick up, taking care of the coal fire and removal of ashes, and having to do it all by hand. Systems were automated by installing stokers (a type of machine designed to burn coal automatically), gas burners, or oil burners. The slow pick up of the system made the use of night set-back thermostats impractical. It took too long to get the house back up to the daytime temperature. It really made no sense to lower the heat at night. What were the answers to these problems? Return to steam heating? Never! Problems are made to be solved so this is what was done.

FORCED CIRCULATION HOT WATER HEATING SYSTEMS

Gravity systems were converted to forced circulation systems by installing the following parts (see Figure 9-3).

1. **A Circulator (4)**—This motor-driven pump is installed on the return line as close to the boiler as possible. The location of the circulator has been a major topic of discussion for many years. We will discuss the installation of the circulator on the supply piping later on in this chapter. The thermostat controls the circulator. On a call for heat the circulator turns on, forcing water from the return into the bottom of the boiler. This pushes the hot water from the boiler into the supply main and then into the radiation. When the room warms up and the thermostat is satisfied, the circulator shuts off (see Figure 9-4).

2. **Flow Control Valve (5)**—This valve is installed in the supply main, as close to the boiler as possible. The purpose of this valve is to keep the heated water in the boiler and not allow gravity circulation. The circulator forces the flow control valve open when there is a call for heat. Every flow control valve has a device that will allow the valve to be left in the open position. This is done so that if the circulator fails, the system can still heat the building by gravity circulation (see Figure 9-5).

3. **Expansion Tank (6)**—The open expansion tanks, located in a closet or the attic, cannot be used with forced circulation because they spill over when the circulator comes on. There is still a need for an expansion space in the system because as water is heated, it does expand. If there is no place for this additional volume of water to go, it will raise the pressure in the system and open the relief valve. The expansion tank is located in the boiler room, close to the boiler, and usually suspended from the ceiling. A connection is made from the boiler or the flow control valve to the bottom of the tank. As the system is filled with water, some of the water will enter the tank. An air bubble is formed in the tank because there is no way for the air to get out. This

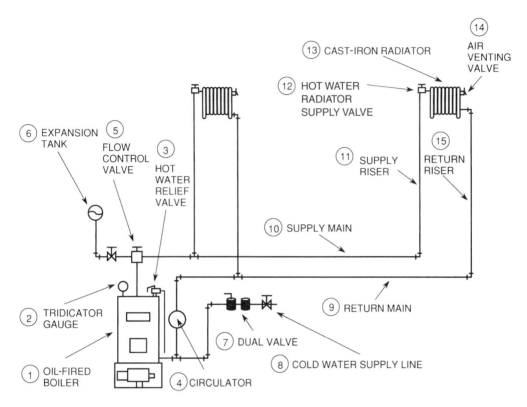

Figure 9-3 **Forced circulation hot water heating system.**

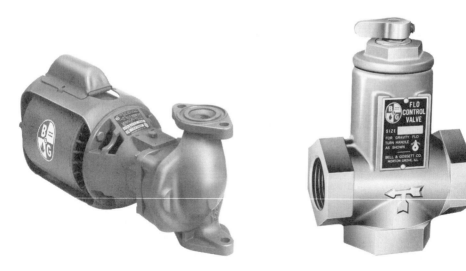

Figure 9-4 **How water system circulator.**
Courtesy: Bell and Gossett.

Figure 9-5 **Hot water system flow control valve prevents gravity circulation of heated boiler water.** *Courtesy: Bell and Gossett.*

air bubble allows the water in the system to expand without raising the system pressure (see Figure 9-6).

4. **The Dual Valve (7)**—This valve is used to maintain the proper amount of water pressure to keep the system full of water. The first part of the valve is a pressure-reducing valve that reduces the cold water pressure to the pressure required to keep

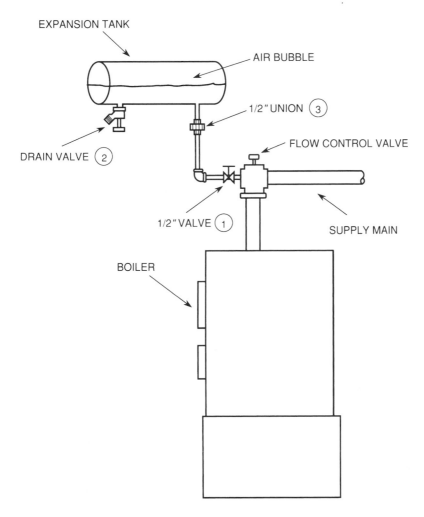

TO ESTABLISH AN AIR BUBBLE IN THE TANK:
CLOSE VALVE (1)
OPEN DRAIN VALVE (2)
CRACK 1/2" UNION TO ALLOW AIR TO ENTER THE TANK (3)
REMOVE 2 BUCKETS OF WATER, THEN CLOSE THE DRAIN VALVE AND TIGHTEN THE UNION.
OPEN VALVE (1) TO RESUME NORMAL OPERATION

Figure 9-6 **Procedure for establishing an air bubble in an expansion tank.**

the system full. For every foot of altitude in the system approximately 1/2 psi of water pressure is required. The second part of the valve is a relief valve that is set to open at 30 psi. Some of the valves have a bypass built in so that the system can be filled quickly. These bypass valves must be closed once the system is full to prevent the full water pressure, usually 55 psi, from entering the system. This would have water continuously running out of the relief valve.

5. **The Tridicator Gauge (2)**—This gauge, also known as a thermaltimeter gauge, is installed in the boiler and indicates the water temperature, the pressure in the system, and the altitude of water in the system. The altitude gauge is set by hand when the system is being filled. Once the system is filled with cold water, you set the altitude gauge indicator to correspond with the pressure gauge indicator. The pressure indicator will move to the right of the altitude indicator when the water is heated. Just looking at the gauge when the boiler is operating will allow you to tell if the system is full of water or not. Remember that all the radiators must be full of water so that the water will circulate through and heat them (see Figure 9-7).

Operation of the Forced Circulation Hot Water System Follows (See Figure 9-3):

1. The water in the boiler (1) is kept hot, 180°F, and does not circulate by gravity because the flow control valve (5) keeps it in the boiler.
2. The thermostat calls for heat, turning the circulator (4) on, circulating hot water from the boiler into the system.

Figure 9-7 **Thermaltimeter gauge on hot water system boiler indicates water temperature, system pressure, and height (altitude) of water in the system.** *Courtesy: Burnham Corporation.*

3. Cold water returning from the system enters the boiler where it is heated. There is a control in the boiler that will shut the circulator off if the water temperature is too cold to heat the rooms. This allows the boiler to catch up and heat the water.

4. Once the thermostat is satisfied, the circulator will shut off, but the boiler will continue to run to heat the water in it to be ready for the next heating cycle.

HYDRONIC HEATING SYSTEMS

The forced circulation systems described before were really gravity supply systems that were made to operate automatically. Some of the old problems were still with us. The main problem was the piping, which took up too much room in the basement. The large radiators were eyesores and had to be replaced. The evolution of hot water heating systems led to the hydronic systems, which are being installed in many new residential buildings. Hydronic systems are forced circulation systems that have smaller boilers designed specifically for hot water heating (see Figure 9-8). They may not have an internal steam chest,

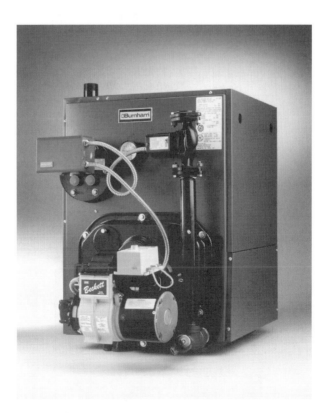

Figure 9-8 **Oil-fired Burnham hot water system boiler. Complete package includes oil burner, circulator, and combination hot water control and circulator relay.** *Courtesy: Burnham Corporation.*

which would reduce the internal volume of the boiler. The use of small boilers, baseboard radiation, and smaller supply and return piping, results in greatly reduced water capacity in the system. Heating the small amount of water leads to the fuel economy that is so essential to the average homeowner.

Hydronic systems may be piped in a series loop system, one-pipe-mono-flow tee system, two-pipe-reverse return system, or the two-pipe-direct return system (see Figure 9-9). The one-pipe-mono-flow tee system reduces the amount of piping in the building by eliminating the return pipe hook-up to the radiators. The mono-flow, or diverter tees are installed as shown in Figure 9-10. The fittings are installed in the supply main and direct the water to the radiation. There is no longer a need for a return main. Using smaller pipe, usually copper tubing, allows us to install these systems without clogging the basement

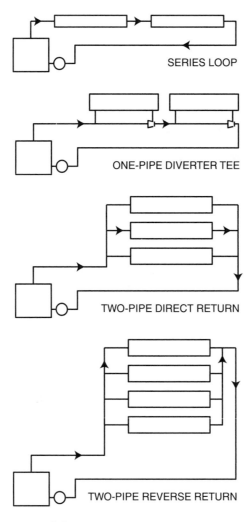

Figure 9-9 **Hydronic system piping arrangements.**

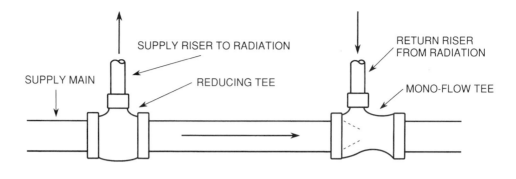

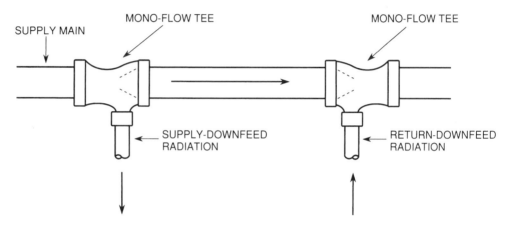

Figure 9-10   **Piping hook-up for mono-flow (diverter tees). Tees must be set apart the full length of the radiation.**

of the house with pipe. You can hide all this piping in the walls or ceiling and thus have the use of the entire basement for living space.

Hydronic Heating System Parts Follow (See Figure 9-11):

1. Air eliminators are installed either in the boiler outlet (see Figure 9-12) or on the supply piping close to the boiler (see Figures 9-13). When the piping type is used, an automatic air venting valve is installed in the top (see Figure 9-14). The other outlet on the air eliminator is for the connection to the expansion tank. Air eliminators are necessary to remove air bubbles from the small size piping. These bubbles create annoying noises and can collect and cut off circulation to parts of the system.
2. The expansion tanks used for the hydronic heating systems are of the diaphragm type (see Figure 9-15). They have a separate chamber where the system water collects when the water is heated causing expansion. This chamber is separated from the rest of the tank by a neoprene diaphragm. On the other side of the diaphragm is a sealed,

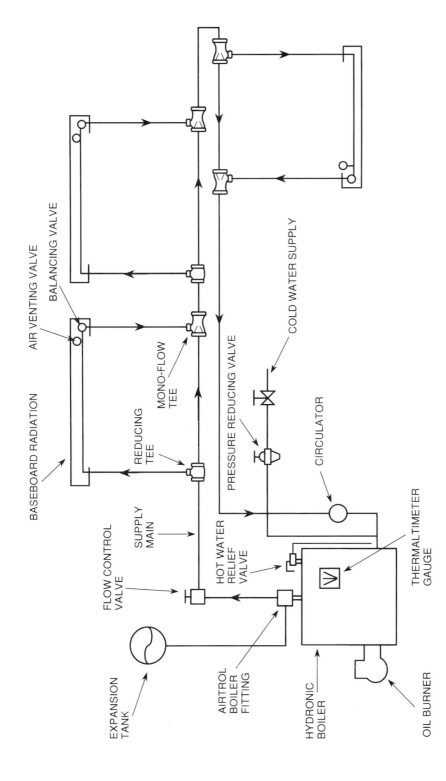

BASEBOARD RADIATION

AIR VENTING VALVE

BALANCING VALVE

MONO-FLOW TEE

REDUCING TEE

PRESSURE REDUCING VALVE

COLD WATER SUPPLY

FLOW CONTROL VALVE

SUPPLY MAIN

HOT WATER RELIEF VALVE

CIRCULATOR

THERMALTIMETER GAUGE

EXPANSION TANK

AIRTROL BOILER FITTING

HYDRONIC BOILER

OIL BURNER

Figure 9-11 **Single zone hydronic heating system with mono-flow tees.**

Figure 9-12 Airtrol air eliminator is installed directly into the boiler water outlet. This fitting diverts air entrained in the boiler water to the expansion tank. *Courtesy: Bell and Gossett.*

Figure 9-13 Air eliminator is installed on supply piping 18″ horizontal run from boiler outlet. *Courtesy: Watts Regulator Company.*

Figure 9-14 Automatic air venting valve for hydronic heating system. *Courtesy: Watts Regulator Company.*

Figure 9-15 Air eliminator with air vent and diaphragm-type expansion tank on supply main. *Courtesy: Amtrol, Inc.*

pressurized air chamber. When expansion takes place, the diaphragm is forced into the air chamber, which compresses. This allows the system water to expand without raising the system pressure. The advantage this type of tank has over the single chamber type is that the air cannot be absorbed by the water, which eliminates the air chamber. These tanks do not have to be drained to reestablish the air bubble they have to contain.

3. Boiler feed valves are of the pressure reducing type as shown in Figures 9-16 and 9-17. There are some local requirements for a back-flow preventer in the boiler feed line (see Figure 9-18). These valves eliminate the possibility of boiler water contaminating the potable water supply. Check your local building code to find out if there is such a requirement.

4. The hot water relief valves are separate from the feed valves (see Figure 9-19). Many localities require this so that the older dual-type valves can no longer be used.

5. The flow control valves are reduced in size and made so that copper pipe can be soldered directly to the valve, which simplifies their installation (see Figure 9-20).

6. The circulators have been greatly reduced in size and feature quiet operation and low operating cost (see Figure 9-21).

7. The use of baseboard radiation along the outer walls of the house will assure that the house is heated evenly.

Circulator Location

The development of hydronic heating systems has taken place over a period of more than 50 years. The changeover from gravity supply systems to forced circulation systems required the addition of the circulator. Early circulators were large and heavy. The easiest place to mount them was on the return piping close to the boiler, which could be done with the least amount of disruption to the existing pipework. So many circulators were installed this way, the practice became traditional. Occasionally there would be a job in which the top floor radiators had to have the air bled out of them every couple of days. Nobody thought the circulator was the cause of the problem, but it was.

A circulator is not a traditional pump. It moves the water in the closed piping system by creating differences in water pressure. Figure 9-22 illustrates a single-loop hydronic system. The air eliminator and expansion tank are on the supply piping near the boiler. The pressure reducing feed valve, which is connected to the piping at the expansion tank, is set at 12 psi, maintaining that pressure throughout the system, which is the normal operating pressure for a residential hydronic system. When there is a call for heat the circulator starts, the pressure in the supply and return lines becomes unbalanced, and the water starts to circulate through the system. There is only one place, known as *the point of no pressure change*, where the pressure remains constant and that is where the expansion tank is connected to the piping.

In order to change the pressure in the expansion tank, we would have to either add water to it or take water out of it. The circulator can do neither of these things. Because this is true, the pressure remains constant at this point in the piping. When the circulator

Figure 9-16 Hot water system reducing valve maintains system pressure to keep radiation full of water. *Courtesy: Bell and Gossett.*

Figure 9-17 Hot water feed reducing valve for large systems. *Courtesy: Watts Regulator Company.*

Figure 9-18 The back-flow preventer is required by some local building codes. This unit is installed between the pressure reducing valve and the cold water line. The purpose is to prevent contamination of the potable water if boiler water were to back into the cold water piping. *Courtesy: Watts Regulator Company.*

Figure 9-19 Relief valve for hot water boiler. *Courtesy: Watts Regulator Company.*

Figure 9-20 Flow control valve can be soldered directly to system copper piping. *Courtesy: Watts Regulator Company.*

Figure 9-21 Hydronic system circulator. *Courtesy: Bell and Gossett.*

is mounted on the return line, it is moving water in the direction of the point of no pressure change. In order to get the water to move, it has to do one of two things. It either has to overcome the 12 psi pressure at the expansion tank or reduce the pressure in the return piping. Since most of the small circulators now in use on residential systems can only build up a head pressure of approximately 6 psi, there is no way it can overcome the pressure at the expansion tank. What happens is that it reduces the pressure in the return line by the same 6 psi. Now the higher pressure water in the boiler can move through the supply line into the radiation because high pressure always goes to low pressure. The system seems to be operating properly, but there is a problem.

Water absorbs air. There is always air entrained in the water circulating through the system. The normal operating pressure keeps the air entrained in the water. Reducing the pressure allows some of the air to escape from the water and form bubbles in the piping. You can notice a similar reaction when you pop the lid on a can of soda. The sudden release of pressure on the inside of the can allows the gas in the liquid—carbon dioxide— to escape. If you leave the can open, eventually all of the gas will escape from the liquid. If we are lucky, the bubbles of air that form inside the system piping will be carried along by the water through the air eliminator and released to the atmosphere. If we are not lucky, the air will collect in one of the radiators—usually on the top floor of the building— where it has to be removed to allow the radiator to heat up.

Figure 9-23 shows a similar single-loop system with the circulator mounted on the supply line. The circulator will be pumping away from the point of no pressure change. This allows the circulator to add its operating head pressure to the system pressure. The

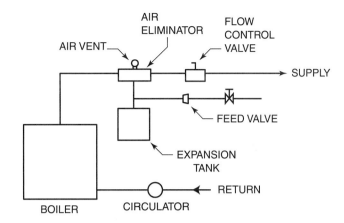

Figure 9-22 **Single-loop hydronic system with circulator on return piping.**

unbalance in the piping pressure created will move the water from the higher pressure supply side to lower pressure return side piping. Increasing the operating pressure will lock entrained air in the system water and eliminate the air bubble problem.

All of this has been common knowledge for many years. You might ask why manufacturers are still sending out packaged hydronic boilers with the circulators mounted on the return side of the boiler. The answer is that it is easier and more convenient for them. The circulator can be prewired and the entire unit packed into a compact crate for easy shipping and storage. You can see in Figure 9-23 that the piping on the supply side of the boiler is quite complicated. Add a couple of zones to the system and it becomes cumbersome and requires the construction of supporting racks to keep everything in place. However, it is worth the extra effort because a service call on a cold night that requires bleeding air out of a radiator can be avoided.

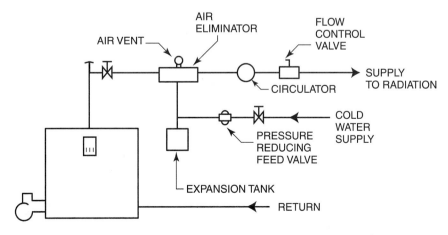

Figure 9-23 **Single-loop hydronic system with circulator on supply piping.**

RADIANT HEATING SYSTEMS

Radiant heating systems are not designed to replace the heat lost from a building to the outside air. They are designed to replace the heat lost from your body. Our bodies are actually very efficient heating units. We produce a great deal more heat than we need. The excess heat is removed by evaporation when we perspire, by convection when air moves around us, and by radiation because heat always goes to cold. If we stand near something warmer than we are, we feel warm because heat radiating from the warmer surface will be absorbed by our cooler body. If we stand near something colder than we are, we feel cold because heat from our body will radiate to the cooler surface. Our average external temperature when we are dressed is approximately 85°F. If we are surrounded by objects that are heated to a temperature in the range of 85 to 95°F we will not lose any of our body heat and will feel quite comfortable regardless of the actual room temperature, which could be considerably lower. That is the basic idea behind radiant heating.

To establish the desired environment, we have to heat a large surface—either the floor, the ceiling, or the walls—to our body temperature of 85°F. The most popular setting for water heated systems is in the floor. Electrical radiant systems are usually installed in the walls or ceiling. In new buildings, the floor systems consist of heat-conducting plastic or copper coils embedded in a well-insulated concrete slab (see Figure 9-24). When a radiant system is installed in an existing building, the coils of piping are stapled to the underside of the flooring and then insulated (see Figure 9-25). The temperature of the water circulated through the system depends on the flooring material. For a concrete floor we would need approximately 120°F and for a wood floor in which the tubing is stapled below it, we would need approximately 140°F.

Heating with low temperature water presents a serious problem. Flue gasses will condense in the flue gas passages of the boiler if the water temperature is below 140°F. The flue gas contains traces of sulfur, which becomes sulfuric acid when the gas condenses. The acid will attack the boiler and the flue pipe connection to the chimney. The

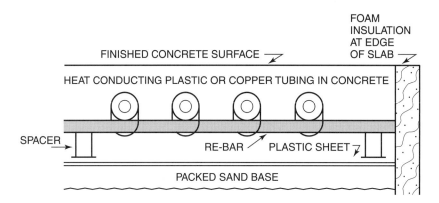

Figure 9-24 **Cross section of radiant heating coil embedded in concrete slab.**

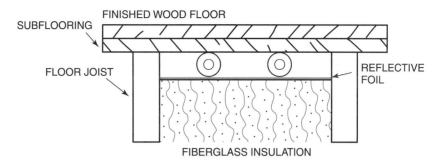

Figure 9-25 Radiant heating coil stapled to underside of subflooring.

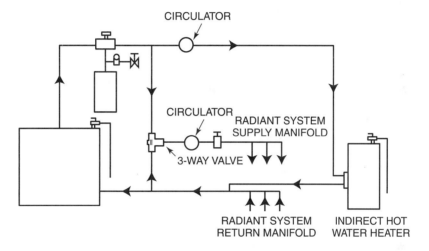

Figure 9-26 Three-way valve controls water temperature to radiant heating grid. Hot water heater needs 180°F water, radiant grid needs 120°F water.

only way to avoid this is to maintain a boiler temperature of at least 160°F. Figure 9-26 illustrates a boiler hook-up for a radiant system that has a 3-way valve that can be set to the desired system temperature (see Figure 9-27). You can see that the hot water from the boiler is mixed with the cooler water in the heating loop and that will solve the problem. This piping set-up also allows the boiler to generate domestic hot water.

ZONING

Another popular modification of systems is zoning. This means that separate parts of a building can be heated individually from the same boiler. Zoning can be accomplished by installing a separate heating loop starting at the boiler with a flow control valve into a single loop and a circulator that is controlled by a thermostat in the separate zone (see Figure 9-28). Another method is to use zone valves on the supply piping for separate

Figure 9-27 **Three-way valve, circulator and supply, and return manifolds for radiant floor heating zone.**

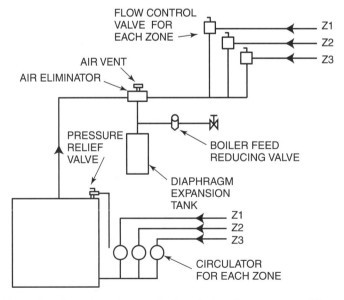

Figure 9-28 **Piping hook-up for a 3-zone hydronic heating system with separate circulators for each zone.**

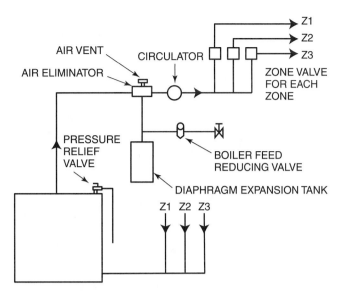

Figure 9-29 Piping hook-up for a 3-zone hydronic heating system with zone valves. Installing zone valves on the supply line to each zone eliminates the need for flow control valves.

parts of the building (see Figures 9-29 and 9-30). The thermostat controls the zone valve, which opens on demand for heat. As the valve opens, a switch inside the valve turns the circulator on. As the other zone valves are closed, the water will only circulate through the radiation in the zone that is calling for heat. The circulators can be installed on the return manifold at the boiler (see Figure 9-31). The piping arrangement shown allows individual purging of each zone to remove all air pockets.

COMBINATION SYSTEMS

It is beyond the scope of this text to attempt to illustrate the possible variations in hydronic heating systems. The system described here is fairly common in residential buildings. We have a 3-zone system, a basic baseboard loop in part of the building, a radiant floor system in another section, and an indirect hot water heater all connected to a single boiler. Two of the zones, the baseboard loop and the indirect water heater require water temperatures of 180°F. The radiant section requires a water temperature of 120°F. To accomplish this we can utilize a primary and secondary piping loop piping system.

The primary loop (see Figure 9-32) should be done in the full size of the boiler outlet in either black pipe or copper tubing. Starting from the boiler there is a tee for a probe-type low water cutoff. This control is necessary on any hydronic system in which part of the piping is below the level of the boiler. We progress from that tee to a second tee for our boiler purge valve and main line connection. A full size ball or gate valve is installed

Figure 9-30   **Zone valves on supply main.**

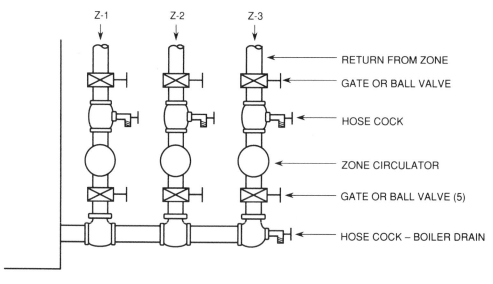

Figure 9-31   **Return manifold piping for a 3-zone circulator system. Each zone can be purged by closing valve 5 and opening the drain valve above the circulator. Any air trapped in the zone piping will have to come out of the drain valve. Once a steady stream of water is established, valve 5 must be opened for normal operation.**

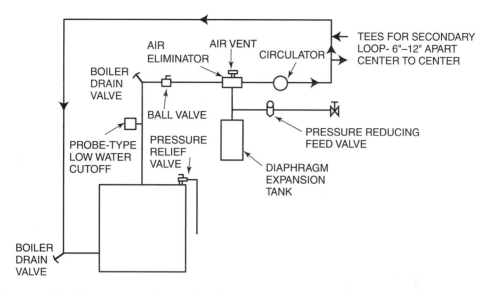

Figure 9-32   **Primary loop piping. Use pipe the same size as the boiler outlet.**

before the air eliminator. This valve is used when the system is filled, as will be explained later. The air eliminator and diaphragm-type expansion tank are next. There should be a horizontal run of at least 18″ before the air eliminator. The boiler feed valve is connected into the line between the air eliminator and the expansion tank. Remember, this is the point of no pressure change so the reducing valve we use for a boiler feed will not be affected by fluctuations in pressure caused by the circulator. The circulator is next and then the two tees where the secondary loop is connected. These tees should be installed 6″ to no more than 12″ apart center to center. From the second tee, we return to the return connection of the boiler. Use a reducing tee for a boiler drain at the boiler.

The secondary loop starts at the tees on the primary loop (see Figure 9-32). Since there are 3 zones we must have a supply manifold and a matching return manifold. The zones are piped out of the manifold with the circulators on the supply side. The radiant floor zone has a preset 3-way valve to control the water temperature in that loop.

Combination System Operation (see Figure 9-33):

1. The boiler water temperature is maintained at 180°F.
2. When one of the zones calls for heat, the primary circulator starts and sends 180°F water through the primary loop piping.
3. The zone circulator starts and pumps the water from the primary loop into the zone loop.
4. When the temperature of the water in the boiler drops to 170°F, the burner starts to reheat the water.

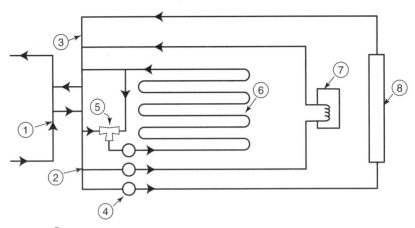

① PRIMARY LOOP
② SUPPLY MANIFOLD–SECONDARY LOOP
③ RETURN MANIFOLD–SECONDARY LOOP
④ ZONE CIRCULATORS
⑤ 3-WAY VALVE FOR RADIANT FLOOR LOOP
⑥ RADIANT FLOOR LOOP—120°F
⑦ INDIRECT HOT WATER HEATER LOOP—180°F
⑧ BASEBOARD HEATING LOOP—180°F

Figure 9-33   **Secondary loop piping for 3-zone hydronic system when zones have different water temperature requirements.**

5. When the call for heat is satisfied, the zone circulator and the primary loop circulator will stop. The burner will continue to operate until the water temperature in the boiler is restored to 180°F.

   To fill the system, close the ball valve on the primary loop supply line. Open the valve in the water feed line and attach a hose to the boiler drain valve on the primary loop. Allow the water to run into the system until a steady flow of water with no air pockets runs out of the hose. Once this happens you can be certain that the boiler and all of the zones are full of water and the system is ready to operate.

COLD START SYSTEMS

There is a continuing effort by the industry to raise the level of efficiency of oil-fired heating equipment. One of the devices recently adopted is the cold start system. The idea is to avoid stand-by loss of heat after a heating cycle is completed. The burner and circulator start together when there is a call for heat from a heating zone or the hot water heater. At the end of the cycle, any remaining hot water in the boiler is circulated through the hot water heater where the heat is collected. This cools the boiler water and eliminates any stand-by loss of heat either up the chimney or through the boiler insulation.

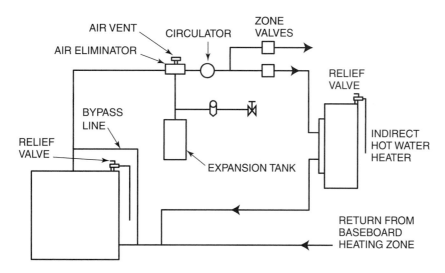

Figure 9-34 **Piping hook-up for cold start system includes bypass line to help boiler reach operating temperature quickly and minimize the formation of condensation in the boiler.**

The problem of condensation of the flue gas cannot be ignored. Every heating cycle will create condensation until the water temperature in the boiler reaches 140°F. This is like painting the inside of the boiler with sulfuric acid a couple of times every day. There are boilers built with a protective coating in the flue gas passages that can handle the acid. There are other boilers that contain very little water. They heat up quickly so the damage is minimized. There is a special piping hook-up (see Figure 9-34) in which a bypass line is connected between the supply and return lines close to the boiler. This allows some of the heated boiler water to circulate back into the boiler, which helps to raise the water temperature quickly, reducing the amount of condensation formed.

To Briefly Summarize

1. Hot water heating systems solved many problems inherent in steam systems.
2. The hot water system must be full of water to operate properly.
3. The hydronic systems features small boilers, small size piping, and baseboard radiation. These systems are being installed in many new homes. They have achieved great popularity because they easily adapt to any building situation.
4. An alternative to baseboard radiation systems are the radiant systems. These have a grid of heating piping embedded in the floor and supply even heat throughout the building.
5. There are many variations in hot water heating systems. This is because of the evolutionary process that has taken place over a long period of time.
6. New equipment and ideas take a long time to filter into the buildings you will be working in. You can expect to see many old systems still operating and providing a comfortable environment for the people living with them.

Please Answer These Questions

1. Explain the function of the following parts:
   A. The dual valve
   B. The thermaltimeter gauge
   C. The expansion tank
   D. The circulator
   E. The flow control valve
2. Explain how you would fill a hydronic system with water.
3. Why is the radiation for a hot water system larger than the radiation for a steam system in the same size building?
4. Why is it less expensive to operate a hot water system as compared to a steam system for the same building?
5. Explain how a mono-flow tee directs water to the radiation.
6. Explain the function of the air eliminator.
7. Why would a water filled expansion tank cause the relief valve to open?
8. What does *zoning* mean and how can it be accomplished?
9. What is a radiant heating system?
10. When is it necessary to install a low water cutoff on a hot water heating system? What type of control is used for this purpose?
11. What happens when a boiler is set up to "cold start" and why is it a problem?
12. Explain the advantage gained by installing the circulator on the supply side piping.
13. Installing zone valves on the supply piping eliminates the need for flow control valves. Why is this statement true or false?
14. Why should the boiler feed pressure reducing valve be connected to the piping at the expansion tank connection to the air eliminator?
15. Explain the function of the 3-way valve in a radiant floor heating system.

CHAPTER

10

WARM AIR HEATING SYSTEMS

Warm air heating systems evolved from the wood or coal burning unit that provided the heat and possibly the cooking stove for a home. These systems have retained their popularity for the following reasons:

1. They are inexpensive to operate because the air is heated directly in the furnace without having to heat water or produce steam to heat the air in the building.
2. There are no radiators, which take up a considerable amount of wall space. The supply and return duct registers fit flat against the wall or on the floor.
3. There is never any danger of freeze up, which would crack a steam or hot water system boiler, because there is no water to freeze. This makes warm air systems ideal for vacation type homes that are not in use continuously.
4. Modern units can be adapted to cool as well as heat the building.

GRAVITY FEED WARM AIR SYSTEM

As buildings became more sophisticated, the warm air furnace was developed. Installing warm air systems marked the beginning of the mechanical heating and ventilating industry that separated heating from plumbing. The furnace was installed in the basement of the building and the system was connected to the house with a supply duct and a return duct. The air that was being heated in the furnace would then rise into the house through the supply duct while the cooler air in the house would circulate to the furnace through the return duct, where it was heated. This is known as a gravity feed system since there is no fan involved to force the air through the ductwork into the building. The dominant feature of this type of system was the large duct piping that took up the entire basement of the house. Each room had a set of supply and return ducts (see Figure 10-1).

The furnace had two major parts, the heat exchanger and the plenum. The heat exchanger was made of cast iron and this is where the fuel was burned. In the original systems the fuel used was either coal or wood. The heat exchanger had to be strong enough to withstand the heat of the fire without cracking. The plenum was a sheet metal casing that surrounded the heat exchanger. The supply ducts were connected to the top of the

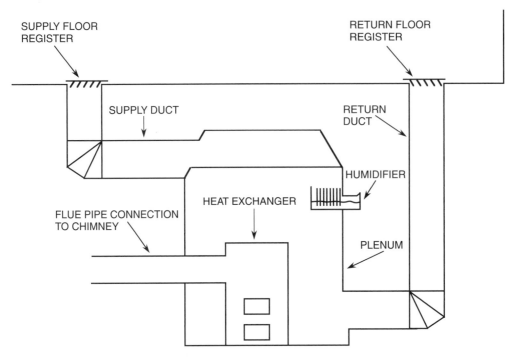

Figure 10-1 **Gravity feed warm air system.**

plenum and the return ducts were connected to the bottom of the plenum. When a fire is started in the heat exchanger, it gets quite hot. This heats the air in the plenum, which then rises into the building, and the circulation begins. This is the least expensive way to heat a house because we are heating the air directly and not using water or steam to do the job for us. One of the problems with this system is that heating the air directly will drive all the moisture out of it. To replace moisture in the air, a device called a humidifier was developed to be installed in the plenum. The original humidifier was merely a flat tray that held about a quart of water. This tray would be filled daily and as the air was heated, the water would evaporate and add moisture to the air.

Oil burners were first installed in warm air furnaces in the 1930s. These early machines were quite noisy and as the sound carried quite well through the ducts it could be heard all over the house. But it was still better than shoveling coal and cleaning the ashes out every day. Once electrical wiring was attached to the unit, someone figured out a way to install a large fan that would force the air through the ducts to speed up the heating process. You can imagine the dust flying when that fan was turned on for the first time. Filters were installed to control the dust and as each problem was solved these systems became more popular.

FORCED CIRCULATION WARM AIR SYSTEM

The building boom that followed the end of World War II brought with it a large influx of warm air heating systems. Building contractors were able to save a substantial amount of money on each heating system because these systems were much less expensive to install than either steam or hot water. The furnaces and ductwork were made smaller and the buildings were designed so that the ductwork would fit inside the walls and under the floors. Each furnace came with a blower section attached (see Figure 10-2). There was a filter built into the unit where the return duct was connected. This effectively kept dust out of the plenum and did an effective job of dust control in the building. The humidifiers were improved and each unit was equipped with one from the factory (see Figure 10-3).

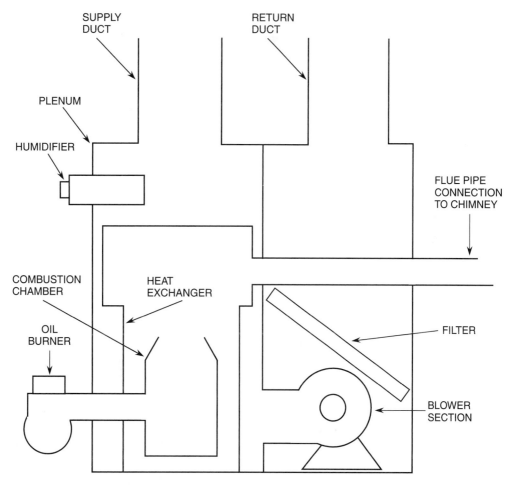

Figure 10-2 **Forced circulation warm air system.**

Figure 10-3   **Vapomax humidifier mounts in the furnace plenum or supply duct. This unit contains an electric heater that vaporizes water to add humidity to the system air.** *Courtesy: Adams Manufacturing Company.*

A simple electrical control system was developed to keep the unit operating safely and efficiently.

The control system includes a thermostat located in the living space. There is a double switch, the fan–limit switch mounted in the plenum, and a combustion safety control.

The Sequence of Operation of a Forced Circulation Warm Air Heating System Is as Follows:

1. The thermostat calls for heat and the burner starts.
2. The heat exchanger gets hot and heats the air in the plenum.
3. The fan switch feels the heat and turns the fan on when the temperature reaches approximately 120°F.
4. Warm air is now forced through the supply ducts into the building. Cooler air from the building returns to the furnace through the return duct, passes through the filter, and enters the plenum where it is heated. If during this cycle the temperature of air in the plenum drops below 100°F, the fan switch will shut the fan off. The burner will continue to run until the thermostat is satisfied, then it shuts off. The fan will continue to operate as long as the air in the plenum is hot.

MODERN WARM AIR FURNACE

The modern warm air furnace features a high grade steel heat exchanger that is much smaller than the original heat exchanger (see Figure 10-4). A combustion chamber made of ceramic fiber material is installed to protect it (see Figure 10-5). The burners used are flame retention burners that can deliver hot, high efficiency flames. The plenum has been streamlined and looks like a steel cabinet from the outside. Built into this compact unit is the fan or, as it is sometimes referred to, the blower section (see Figure 10-6). The fan can be driven by a direct drive motor or by a system of pulleys and a belt. The pulleys and belt allow us to control the speed of the fan, which is helpful on jobs where noise becomes a source of irritation for the owners of the building. Slowing the fan down will reduce noise and possible vibration in the ductwork (see Figure 10-7).

Figure 10-4 **Heat exchanger for high efficiency oil-fired warm air furnace.** *Courtesy: Dornback Furnace and Foundry Company.*

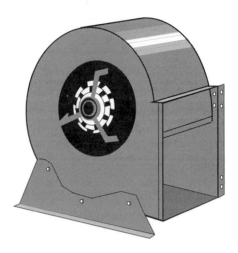

Figure 10-5 **Combustion chamber for high efficiency oil-fired warm air furnace.** *Courtesy: Dornback Furnace and Foundry Company.*

Figure 10-6 **Direct drive, multispeed blower section for warm air furnace.**

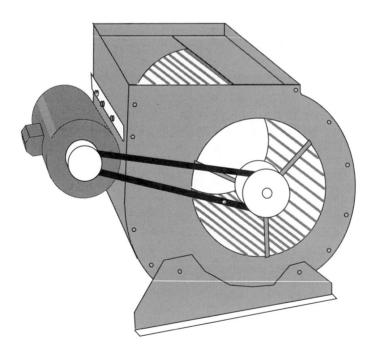

Figure 10-7 **Belt-driven blower section for larger warm air furnaces. Adjustable pulley on the motor allows regulation of fan speed.** *Courtesy: Dornback Furnace and Foundry Company.*

Figure 10-8 shows the Dornback high efficiency condensing oil-fired furnace. Furnaces of this type extract so much heat from the flue gas that the water vapor in the gas condenses. The resulting liquid would corrode the flue pipe, which makes the use of plastic piping necessary for the flue connection of the furnace. This unit is equipped with a draft inducer and an outside air intake duct attached directly to the burner. No chimney is necessary for this unit but it must be installed according to the manufacturer's instructions that are sent with the unit. The unit is also made to fit horizontally in the building when there is no space for the vertical model (see Figure 10-9).

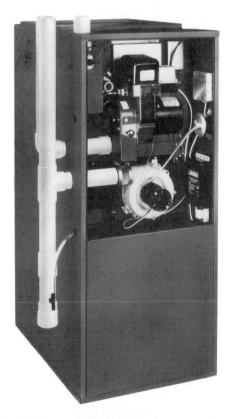

Figure 10-8 **Vertical high efficiency oil-fired warm air system furnace. The unit shown, the Dornback HEO, has a draft inducer and plastic fresh air intake and flue gas exhaust piping. There is a condensate trap on the exhaust piping where liquid can be collected and periodically drained.** *Courtesy: Dornback Furnace and Foundry Company.*

Figure 10-9 **Horizontal high efficiency oil-fired warm air system furnace, the Dornback SHEO.** *Courtesy: Dornback Furnace and Foundry Company.*

AIR FILTERS

The filters are vastly improved and can be either simple fiberglass mats one or two inches thick or electronic devices that capture dust between two electrically charged plates. The filter is always installed where the return air enters the blower section. This is done to ensure that the fan and the plenum remain dust free. If dust is allowed to build up inside the plenum around the heat exchanger, there is a good chance that a fire can start (see Figure 10-10).

HUMIDIFIERS

An entire industry has developed around the humidifier. There are dozens of companies that build them. Each company has its own method and they range from a simple tray with a small float-operated valve to keep it full of water, to a drum turning in a tray of water,

Figure 10-10 **Electronic air cleaner, the Honeywell F-50, is installed where the return air duct enters the furnace.** *Courtesy: Honeywell.*

to a solenoid valve that opens to allow a spray of water to enter the plenum. The problem of dry air is really serious for families that have members with asthma or allergies. You should be aware of this problem even though there is very little that you can do to solve it (see Figure 10-11).

HEATING–COOLING SYSTEM

Perhaps the greatest single advantage of the warm air system is that it can be converted to a heating–cooling system. Most of the furnace manufacturers design the furnaces to accept a refrigeration evaporator coil in the top of the plenum. These coils are built in the shape of a letter A and are called "A" coils (see Figure 10-12). The condensing unit of the refrigeration unit is installed outside the building and connected to the "A" coil by copper tubing that has been precharged with refrigerant (see Figure 10-13). Installing this additional equipment converts the heating system into a true, year-round, climate control system.

Float control is isolated in cold compartment to prevent heat causing liming, clogging.

Removal of motor is simple two-step operation. Unscrew cover plate screw, lift motor from key slots insuring proper alignment.

HUMID-DISC MODEL 1000

A quality constructed humidifier especially built for the competitive market. Rated at 140° F. and 800 fpm. 25 Gallons Per Day moisture output. Installs in ducts as small as 12".

Bottom view shows recessed drain channel (frosted plastic) and oversize plug for rapid draining.

Figure 10-11 Humid Disc Model 1000 humidifier fits into the furnace plenum above the heat exchanger. The rotating discs pick up water from the tank, which evaporates into the system air when the blower is operating. *Courtesy: Adams Manufacturing Company.*

Figure 10-12 Refrigeration evaporator ("A" coil) fits in furnace plenum when unit is used for year-round climate control of building. The "A" coil is connected to the condensing unit shown in Figure 10-13. *Courtesy: Dornback Furnace and Foundry Company.*

Figure 10-13   **The refrigeration condensing unit is installed outside and connected to the "A" coil in the furnace plenum.**   *Courtesy: Dornback Furnace and Foundry Company.*

ZONING

Zoning is made possible by installing motorized dampers in the supply ducts to separate areas of the building. Each of these units has a switch that will turn the burner on for heat or the condensing unit on for cooling when the damper is opened.

KEROSENE SPACE HEATERS

An economical alternative to heating systems are the vented kerosene space heaters developed in Japan (see Figure 10-14). They are ideal replacements for direct electric or heat pump systems that are very expensive to operate. These compact, attractive units are installed in the space to be heated. They have a unique double pipe venting system that allows the air for combustion to be brought in from the outside while the flue gas is vented through the same unit to the outside of the building (see Figure 10-15). The burners used in these units are of the vaporizing pot type and are able to operate at very low GPH rates that can be increased or decreased to match the heating load demand. This enables the unit to run at a variety of settings to deliver the exact amount of heat required to keep the space comfortable. Part of the electronic control system is the programmable clock on the control panel (see Figure 10-16) that can be set for up to 4 separate heating cycles in a 24-hour day.

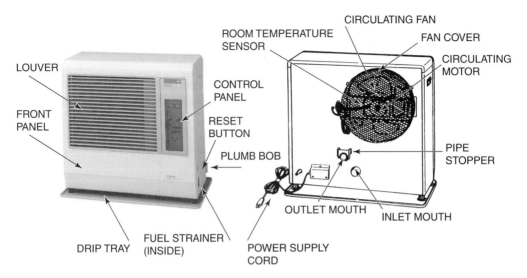

Figure 10-14 **Front and rear view of Toyotomi Laser 73 kerosene space heater.** *Courtesy: Toyotomi U.S.A., Inc.*

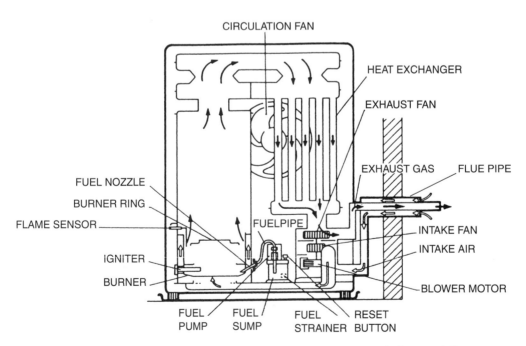

Figure 10-15 **Interior diagram of Toyotomi space heater shows air flow and flue gas exhaust through the unique wall unit.** *Courtesy: Toyotomi U.S.A., Inc.*

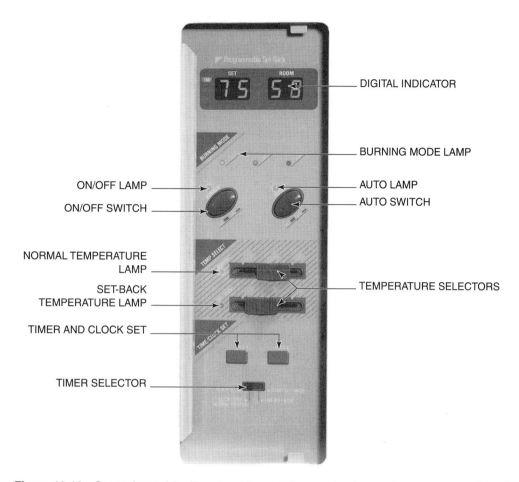

DIGITAL INDICATOR

BURNING MODE LAMP

ON/OFF LAMP

AUTO LAMP

ON/OFF SWITCH

AUTO SWITCH

NORMAL TEMPERATURE
LAMP

SET-BACK
TEMPERATURE LAMP

TEMPERATURE SELECTORS

TIMER AND CLOCK SET

TIMER SELECTOR

Figure 10-16 **Control panel for Toyotomi Laser 73 space heater can be programmed for 4 separate heating cycles a day.** *Courtesy: Toyotomi U.S.A., Inc.*

To operate, the power must be on, the On-Off switch in the On position, the fuel system valves must be open, and the temperature setting on the control panel must be higher than room temperature. First, the blower motor starts sending fresh air through the combustion chamber into the heat exchanger and out through the exhaust pipe. This will clear any residual fumes from the combustion chamber. The prepurge period lasts for 3 minutes and then the ceramic ignitor turns on. The medium burning mode lamp begins to blink, which indicates that fuel is being sent into the combustion chamber. When the flame is established the flame sensor feels it and the light will stop blinking and stay on. The burning mode now changes to the low setting for at least 3 minutes. This slow warm up allows the blower fan to establish the proper air flow through the combustion chamber and heat exchanger. At the end of this part of the cycle, the circulation fan starts and warm air is sent into the space. The burning mode will adjust according to the temperature difference between the setting and the room temperature. If this is the start of a heating cycle and the temperature difference is 10 degrees or more, the unit will go into the high firing mode. The burning mode will change to medium and then low as the temperature difference decreases. When the room temperature reaches the temperature setting, the burner will shut off. The blower will continue to operate to purge the combustion chamber and heat exchanger and the circulation fan will continue until the heat exchanger cools off and then the unit will shut down.

A series of safety devices are built into the system to ensure safe operation. They are electronically connected to the main logic board that supervises the operation continuously. It is very important to attend the training sessions conducted by the manufacturer or an authorized dealer to become competent in servicing and maintaining these units.

There is no way to generate domestic hot water with a warm air furnace. A separate oil, gas, or electric water heater is required. The operation of hot water heaters is discussed in detail in Chapter 17, Domestic Hot Water.

To Briefly Summarize

1. Warm air systems are the most economical to operate because they heat the air directly.
2. Heating the air will remove moisture, which makes a humidifier necessary on every warm air furnace.
3. The air is filtered by installing a filter where the return duct is connected to the furnace.
4. Warm air furnaces can be converted to cooling units by installing a condensing unit outside the building and an "A" coil in the furnace plenum.

Please Answer These Questions

1. What is the function of the humidifier?
2. Where is the filter installed?
3. How is a warm air furnace converted to a heating–cooling unit?
4. What are some of the reasons for installing warm air heating rather than steam or hot water systems?
5. Explain the sequence of operation of a warm air heating system.
6. How is domestic hot water produced in buildings with warm air heating systems?
7. How can zoning of warm air systems be accomplished?
8. Explain how you would adjust the pulleys and belt of the blower section to eliminate noise and vibration in the ducts.

CHAPTER

11

BASIC ELECTRICITY

It is extremely important for oil burner service technicians to have an understanding of electricity. We will cover only the portions of this complex subject that apply to the tasks you have to perform.

The accepted definition of an electrical current is the movement of electrons through a conductor. An electron is a minute, negatively charged particle, that orbits the nucleus of an atom. The electron is held in orbit by the positively charged protons that are found in the nucleus of the atom. Consider a short piece of copper wire. There are millions of copper atoms packed tightly together in the wire. Each atom of copper has 34 electrons in orbit around the nucleus. The movement of the electrons, switching orbit from one nucleus to another, is what holds the atoms together so tightly that the result is a solid piece of wire. When additional electrons are forced into the wire, they speed through to the other end at the speed of light. This is what happens when you turn a switch on to complete a circuit to an appliance. It is important for you to understand that an electrical current represents the movement at the speed of light of millions of electrons through the wiring.

To put electricity to work we must build an electrical circuit. An electrical circuit (See Figure 11-1) is a complete system that must include a source of electrons, a controlling device, a load, and a network of conductors that connects the component parts of the circuit. The conductors are identified as:

1. The feed wire—connects the source of electrons to the controlling device.
2. The section wire—connects the controlling device to the load.
3. The return wire—connects the load back to the source of electrons.

We can compare an electrical circuit to a similar water system. In Figure 11-2, we have a pump supplying water to a sprinkler head. The water from the sprinkler head falls into a tank that is connected to the intake side of the pump. The flow of water to the pump is controlled by a valve. When the valve is open and the pump is in operation, there is a continuous flow of water from the tank into the pump where it is pressurized and forced through the sprinkler head.

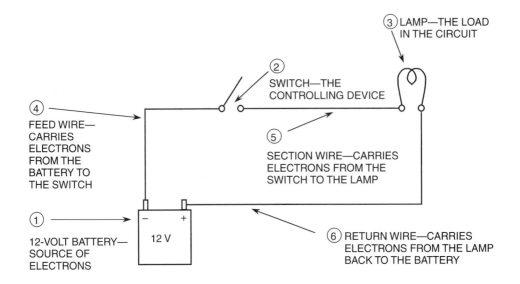

Circuit Operation: When the switch (2) is open the battery (1) forces 12 volts into the feed wire (4) where it waits patiently. When the switch is closed the 12 volts moves through the section wire (5) to the lamp (3) turning it on. The lamp absorbs the volts and the electrons get back to the battery through the return wire (6).

Figure 11-1 **Simple electrical circuit operation.**

The Factors in This System Are as Follows:

1. The amount of water flowing through the system can be expressed in gallons per minute, (GPM).
2. The pressure required to force the water through the sprinkler head can be expressed in pounds per square inch, (psi).
3. The sprinkler head is the load in this water circuit. It absorbs the pressurized water and releases it into the tank at 0 psi.

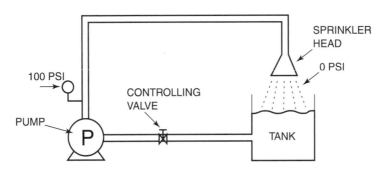

Figure 11-2 **This water circuit is similar to an electrical circuit.**

In an electrical circuit (see Figure 11-1), the pump is replaced by a battery. A battery is a device that releases electrons through a chemical process. The electrons are negatively charged particles that collect on the negative terminal of the battery. There are so many electrons on the terminal they create an electromotive force, or pressure, that we call volts. When the battery is installed in a circuit, the volts become available to carry the electrons to the load. The flow of electrons is controlled by the switch. When the switch is closed, the electrons stream through to the load—in this circuit, a lamp—and turn it on. The load absorbs all the volts and returns the electrons at 0 volts to the battery.

The Factors in an Electrical Circuit Are:

1. The number of electrons flowing through the circuit, usually referred to as the intensity in the circuit, are expressed as amperes (I).
2. The electromotive force, or pressure, in the system is referred to as volts (E).
3. The load, or resistance, in the circuit is referred to as ohms (R).

There is a mathematical relationship between the factors in an electrical circuit that can be expressed by Ohm's Law (see Figure 11-3). If two of the factors are known, the third can be computed by using one of the formulas shown in the diagram. For example:

If the lamp in Figure 11-1 has a resistance of 12 ohms, how many amps will be flowing through this circuit?

$$I = E/R$$

$$I = 12/12$$

$$I = 1 \text{ amp}$$

Every electrical load requires a specific amount of power to operate properly. The unit of electrical power is the watt, (P). We pay for electrical power by the kilowatt (1,000 watts). The formulas for computing watts can be found in Figure 11-4. Here are a few example problems:

How many watts does a 120-volt motor consume if it runs at 4 amps?

$$P = E \times I$$

$$P = 120 \times 4$$

$$P = 480 \text{ watts}$$

$$E = I \times R$$

$$I = \frac{E}{R}$$

$$R = \frac{E}{I}$$

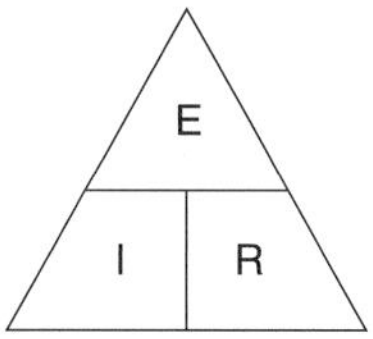

E = VOLTS
I = AMPS
R = OHMS

Figure 11-3 **Ohm's Law triangle and formulas.**

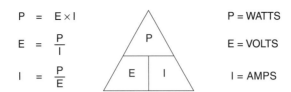

$$P = E \times I$$

$$E = \frac{P}{I}$$

$$I = \frac{P}{E}$$

P = WATTS

E = VOLTS

I = AMPS

Figure 11-4 **Power triangle and formulas.**

What is the resistance of a 120-volt fan motor that consumes 480 watts? (See Figure 11-5.)

We must first compute the amps in the circuit to get the information necessary to determine the resistance of the motor.

$$I = P/E$$

$$I = 480/120$$

$$I = 4 \text{ amps}$$

$$R = E/I$$

$$R = 120/4$$

$$R = 30 \text{ ohms}$$

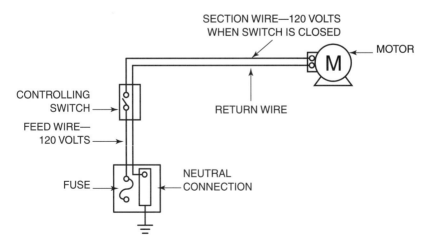

Figure 11-5 **Series circuit has a switch controlling a motor.**

SERIES CIRCUIT

The simplest electrical circuit is the series circuit. The definition of a series circuit is that it has a single path for the electrons to follow. Because this is true, these circuits have some advantages and disadvantages.

The Advantages of Series Circuits Are:

1. **Individual control**. If you want a switch to operate a load, the switch must be wired in series with the load (see Figure 11-6).
2. **Safety controls**. Oil burners have to be supervised by automatic switches known as limit controls. These controls are always wired in series (see Figure 11-7), so that any one of them can shut the oil burner off if an unsafe condition exists. This is true because an opening in any part of the series circuit will shut the circuit down. All switches, manual or automatic, are installed in the feed wire.

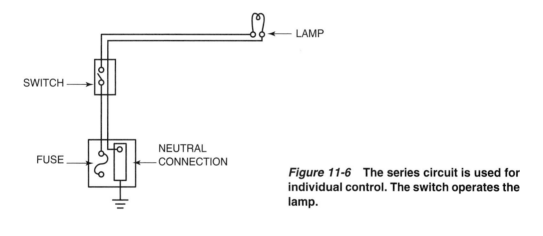

Figure 11-6 **The series circuit is used for individual control. The switch operates the lamp.**

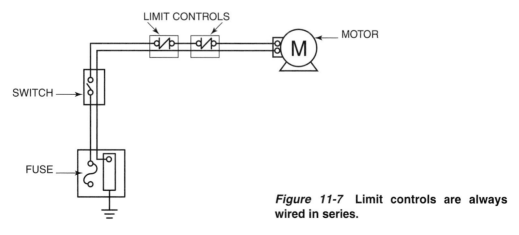

Figure 11-7 **Limit controls are always wired in series.**

Disadvantage of Series Circuits:

1. Loads wired in series have to share the voltage. Because this is true, they do not get the watts they need to operate properly. The available voltage is divided among the loads in direct proportion to their resistance. That means the load with the highest resistance will get the most voltage. We can prove this by using the Ohm's Law formulas.

The diagram in Figure 11-8 shows two 120-volt heaters wired in a series circuit. One has a resistance of 80 ohms and the other has a resistance of 160 ohms. What will happen when the switch is closed? We can find out by following the formulas. First we have to determine the total resistance (Rt) in the circuit. In series circuits the total resistance is the sum of the resistances in the circuit.

$$Rt = R1 + R2$$

$$Rt = 80 + 160$$

$$Rt = 240$$

Now we can compute the amps for this circuit:

$$I = E/Rt$$

$$I = 120/240$$

$$I = .5 \text{ amp}$$

Now we can compute the voltage at each resistance:

$$E = I \times R1 \qquad\qquad E = I \times R2$$

$$E = .5 \times 80 \qquad\qquad E = .5 \times 160$$

$$E = 40 \text{ volts} \qquad\qquad E = 80 \text{ volts}$$

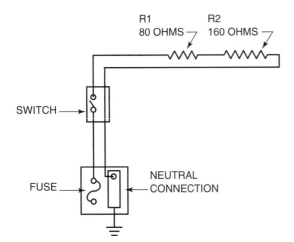

Figure 11-8 **Resistances in series circuits have to share to voltage and do not operate properly.**

Now we can compute the watts rating for each heater. First we have to determine the amps required for each heater.

$$I = E / R1 \qquad\qquad I = E / R2$$

$$I = 120 / 80 \qquad\qquad I = 120 / 160$$

$$I = 1.5 \text{ amps} \qquad\qquad I = .75 \text{ amps}$$

$$P = E \times I \qquad\qquad P = E \times I$$

$$P = 120 \times 1.5 \qquad\qquad P = 120 \times .75$$

$$P = 180 \text{ watts} \qquad\qquad P = 90 \text{ watts}$$

Now we can compute the watts each heater actually received in this circuit:

Heater #1	Heater #2
P = 40 × .5	P = 80 × .5
P = 20 watts	P = 40 watts

Obviously, the heaters will not get hot. Heater #2 might warm up a little, but that will just not work with the loads found in oil burner wiring.

To Briefly Summarize:

1. A series circuit has a single path for the electrons to follow.
2. An opening in any part of a series circuit will shut the circuit down.
3. Safety controls are always wired in series. These controls and any manually operated switches for the circuit are always wired in the feed wire. All of the switches must be closed for the circuit to operate.
4. Loads wired in series will have to share the voltage. The voltage is shared in direct proportion to the resistance of each load. The highest resistance gets the most voltage.

PARALLEL CIRCUITS

A parallel circuit has two or more paths for the electrons to follow. We can use the water system analogy again to explain the operation of a parallel circuit (see Figure 11-9). As you can see from the diagram, the pressure is the same in every branch of the circuit. That is true of the electrical circuit as well. Because of this, when two or more loads are wired in parallel (see Figure 11-10) they each receive the proper voltage and watts they need to operate properly.

The Advantages of Parallel Circuits Are:

1. When two or more switches are wired in parallel, any one of them can operate the circuit. As you can see in Figure 11-11, the switches are each connected to the feed wire

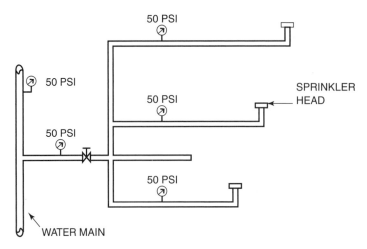

Figure 11-9 Parallel piping to sprinkler system—the water pressure is the same in every branch of the circuit.

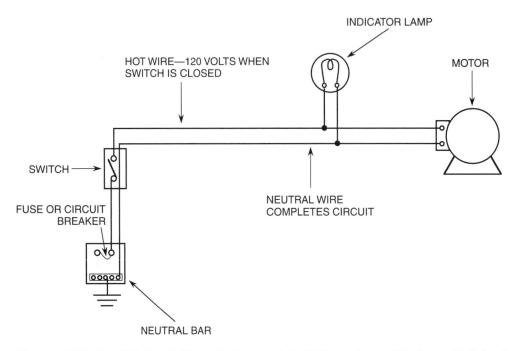

Figure 11-10 Parallel circuit. The switch controls both the motor and the lamp. Both loads operate properly because they get their full voltage.

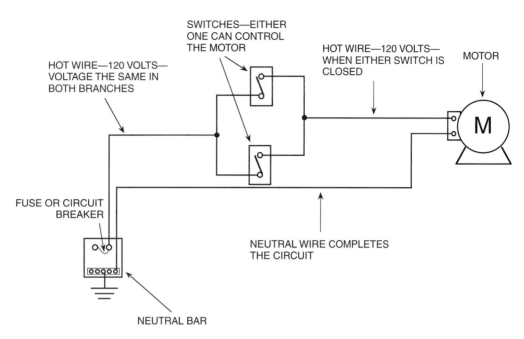

Figure 11-11 **Switches wired in parallel. Either switch can control the motor.**

and section wire. If switch #1 is open and switch #2 is closed, the circuit is completed and the motor operates. If the position of the switches is reversed, the circuit will still operate. If both switches are closed, the circuit will operate as well. Only when both switches are open will the circuit shut down.

2. Loads wired in parallel operate properly because they receive their full voltage. This is true because the voltage is the same in every branch of a parallel circuit. Let us see how the Ohm's Law formulas work with parallel circuits.

In Figure 11-12, we have two heaters wired in parallel. Heater #1 is a 480-watt heater with a resistance of 30 ohms and heater #2 is a 160-watt heater with a resistance of 90 ohms. What will happen when we close the switch?

We can compute the amps in each branch of the circuit:

$$I1 = E/R \qquad I2 = E/R$$

$$I1 = 120/30 \qquad I2 = 120/90$$

$$I1 = 4 \text{ amps} \qquad I2 = 1.33 \text{ amps}$$

$$It = 5.33 \text{ Amps}$$

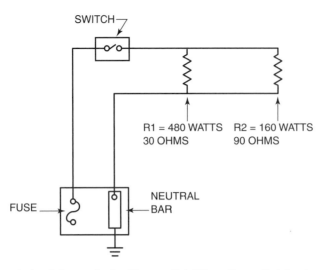

Figure 11-12 **Loads (resistances) wired in parallel. When the switch is closed, both heaters will operate properly because the voltage is the same in every branch of a parallel circuit.**

We can now compute the watts in each branch of the circuit:

$$P1 = E \times I \qquad\qquad P2 = E \times I$$

$$P1 = 120 \times 4 \qquad\qquad P2 = 120 \times 1.33$$

$$P1 = 480 \text{ watts} \qquad P2 = 159.6 \text{ watts}$$

Obviously the heaters receive their proper wattage and will both get hot.

The total amps (It) in a parallel circuit is the sum of the amps in each branch of the circuit. We can prove this with the Ohm's Law formulas but first we have to determine the total resistance (Rt) for the circuit. The formula for Rt in parallel circuits is as follows:

Rt = the reciprocal of the sum of the reciprocals of the resistances in the circuit.

This is not as difficult as it sounds. The reciprocal of a whole number is the number converted into a fraction by making the number 1 the numerator. For example, the reciprocal of the number 9 is 1/9. The reciprocal of a fraction is made by switching the numerator and the denominator. For example, the reciprocal of the fraction 1/4 is 4/1. Using the same diagram (Figure 11-12), we can prove that this method is correct.

$$1/Rt = 1/R1 + 1/R2$$

$$1/Rt = 1/30 + 1/90$$

$$1/Rt = 3/90 + 1/90$$

$$1/Rt = 4/90$$

$$Rt = 90/4$$

$$Rt = 22.5$$

$$I = E/R$$

$$I = 120/22.5$$

$$I = 5.33 \text{ amps}$$

It is interesting to note that the total resistance in this circuit is lower than the smallest resistance in the circuit. This leads to the next question. What would happen if we added another 90-ohm resistance to the circuit? Let us work it out.

$$1/Rt = 1/R1 + 1/R2 + 1/R3$$

$$1/Rt = 1/30 + 1/90 + 1/90$$

$$1/Rt = 3/90 + 1/90 + 1/90$$

$$1/Rt = 5/90$$

$$Rt = 90/5$$

$$Rt = 18 \text{ ohms}$$

$$I = 120/18$$

$$I = 6.67 \text{ amps}$$

Adding a resistance reduced the total resistance, which raised the amps in the circuit, and that is the disadvantage of parallel circuits—they are easy to overload.

To Briefly Summarize:

1. A parallel circuit has two or more paths for the electrons to follow.
2. When two or more switches are wired in parallel any one of them can operate the circuit.
3. The voltage is the same in every branch of parallel circuit.
4. The total amps in the circuit is the sum of the amps in every branch of the circuit.
5. The total resistance in the circuit is the reciprocal of the reciprocals of all of the resistances in the circuit.
6. Parallel circuits are easy to overload.

We can devise a simple oil burner circuit by using both series and parallel circuits together (see Figure 11-13). This combination circuit starts at the fuse. The feed or hot wire, continues in series to the remote control switch, then the burner service switch, then to a limit control, and finally to the flame safeguard control. The two loads in this circuit, the ignition transformer and the oil burner motor, are wired in parallel. When all of the switches are closed, the oil burner will operate. The limit control supervises the operation

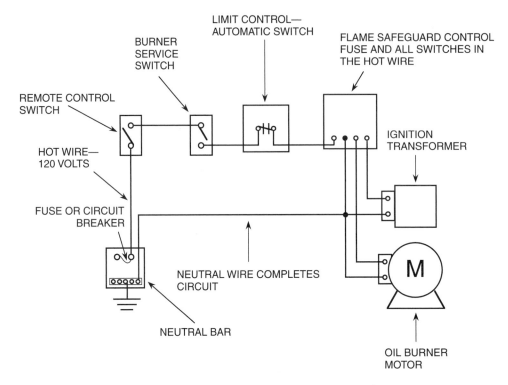

Figure 11-13   **A simple oil burner circuit is a combination of series and parallel circuits.**

of the burner and will shut it off if an unsafe condition exists. Oil burner circuits are always combination circuits.

FUSES–CIRCUIT BREAKERS

The power supply to a building enters through a main electrical panel. From the main panel, the electricity is distributed to the circuits in the building through a series of fuses or circuit breakers. Each circuit is protected by the fuse or circuit breaker. The purpose for these devices is to prevent fires if there is an excessive amount of amperage moving through the circuit. Try to remember that there are millions of electrons moving through the conductor for every ampere in the circuit. If we try to force too many electrons through a conductor, the resulting friction will heat up the conductor and could cause a fire. The size of the wire used in a circuit must be able to handle the amperage without heating up. Domestic oil burner circuits operate at less than 10 amperes. They are normally protected by a 15-ampere fuse or circuit breaker. The wire size used for 15-ampere circuits is #14. Oil burner wiring must conform to the local electrical code.

The word fuse means to melt. The fuses used to protect electrical circuits melt if too many amperes move through the circuit. The fuse is installed in series with the oil burner circuit. All the electricity that goes through this circuit must pass through the fuse first. There are three different types of fuses in use for oil burner circuits. They are the cartridge fuse, the plug fuse, and the fustat. The cartridge fuse fits into the electrical panel on a set of copper clips. Power is fed into the end of the fuse that is closest to the center of the panel and comes out of the end of the fuse that is closest to the outside edge of the panel. The plug fuse fits into a socket that is wired so that the power comes into the bottom of the fuse, moves through the fuse link, and comes out of the side of the fuse. There is a terminal screw on the socket that allows us to connect the fuse to the rest of the circuit. Plug fuses have a glass top. You can clearly see the fusable link inside the fuse. The fustat is an upgraded version of the plug fuse. They must be installed in a base socket that is color coded to accept only the proper amperage fustat. This is done so the homeowner cannot replace a blown fustat with a larger size that might cause a fire or damage to the equipment (see Figure 11-14). The circuit breaker is rapidly phasing out fuses because they

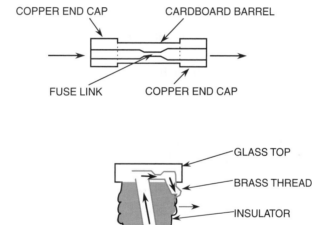

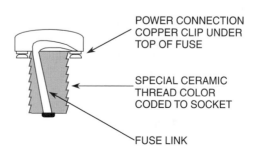

Figure 11-14 **Fuses commonly used for oil burner circuits.**

do not have to be replaced when they shut the circuit off because they act as a switch and only have to be turned back on if they are overloaded and trip out.

The two reasons for fuse blow out, or circuit breaker shut down, are overloads or short circuits. An overload can be caused by a problem with the oil burner motor. Most motors now in use have their own built-in thermal overload protection that has to be manually reset. They will shut off before the fuse can blow or the circuit breaker shuts off. Short circuits are more of a problem than overloads. A short circuit means that there is a connection from a hot wire carrying 120 volts, and either a grounded surface or a neutral wire carrying 0 volts, with no resistance between the wires. Theoretically, 120 amperes could move through the short circuit and that would certainly blow a 15-ampere fuse. Oil burner circuits can develop this kind of problem when the wiring is run too close to a hot surface or when the wiring is damaged by water.

Locating the cause of fuse blow out requires a systematic procedure that starts with connecting a test light to the load terminal of the fuse and then connecting the oil burner circuit to the other leg of the test light. This puts the test light in series with the oil burner circuit with these results (see Figure 11-15):

1. If there is a short circuit the test light will light up because there is a complete circuit for the light.
2. Shut off one of the burner switches. If the light stays on, the short circuit is between the switch and the test light. If the light goes off, the short circuit is farther down the circuit.
3. You can isolate the short circuit by following this procedure.
4. Remember that the short circuit is between the switch that does not shut off the test light and the last switch that did shut the light off.

ELECTRICAL CONNECTIONS

During the course of your work, you will have to install electrical equipment. This work requires making electrical connections either directly to a piece of equipment or splicing wires together to complete a circuit. Remember that electricity moving through a wire is very similar to water running through a pipe. The connections must be tight so the electrons have a solid path through which to flow. If the connections are not tight, a spark will be created every time the circuit comes on. This will burn through the wire until the spark gets big enough to start a fire. You must develop the correct techniques to make solid electrical connections. Then you can do all kinds of electrical work and have confidence in every splice and terminal connection. To be successful with this phase of service work, follow these simple suggestions:

Terminal Screws (see Figure 11-16)

1. Strip approximately 3/4″ of insulation off the wire.
2. Bend the exposed conductor into a hook that can fit around the screw.

1. DISCONNECT WIRE AT FUSE OR CIRCUIT BREAKER.

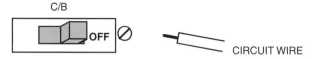

2. CONNECT TEST LIGHT TO C/B TERMINAL AND CIRCUIT WIRE.

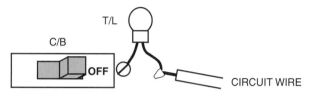

3. TURN C/B ON—LIGHT COMES ON = SHORT CIRCUIT.

4. SHUT OFF NEXT SWITCH IN CIRCUIT—LIGHT SHUTS OFF = SHORT CIRCUIT FURTHER DOWN CIRCUIT.

5. TURN FIRST SWITCH ON AND OPEN SECOND SWITCH—IF LIGHT STAYS ON, SHORT CIRCUIT IS BETWEEN THE SWITCHES.

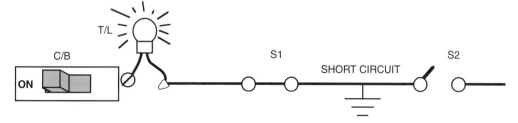

Figure 11-15   **Procedure for locating a short circuit in oil burner wiring.**

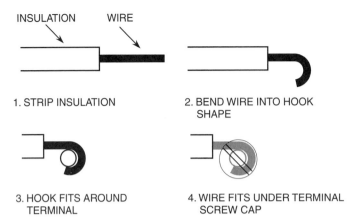

Figure 11-16 **Procedure for connecting a wire to a terminal screw.**

3. Slide the hook over the screw so that the open end of the hook is to the right of the screw.
4. Tighten the screw. As the screw gets tighter, it will close the hook in the wire, making a solid connection.
5. All of the conductor must be under the terminal screw and the insulation should be as close to the screw as possible. Do not leave a tail of conductor sticking out of the screw head or have a stretch of conductor exposed on the insulation side of the terminal screw.

Splices (see Figure 11-17)

1. Strip approximately 3/4" of insulation off each wire that is to be spliced.
2. Hold the wires together so that the exposed ends are next to each other. Hold the wires firmly by the insulation.
3. Twist the exposed wires together in a clockwise direction. Use a pair of pliers and twist the wires as tightly as you can. Do not allow the insulated part of the wires to twist.
4. Screw a wire nut onto the twisted wires and tighten it as tight as you can without allowing the insulated part of the wires to twist. All the exposed wire must fit into the wire nut.

Splicing Stranded and Solid wires (see Figure 11-18)

1. Strip approximately 3/4" of insulation off the solid wire.
2. Strip approximately 1-1/4" of insulation off the stranded wire.
3. Wrap the exposed stranded wire around the exposed solid wire in a clockwise direction.
4. Screw a wire nut onto the splice and make it as tight as you can without allowing the insulated part of the wires to twist. All the exposed wire must fit into the wire nut.

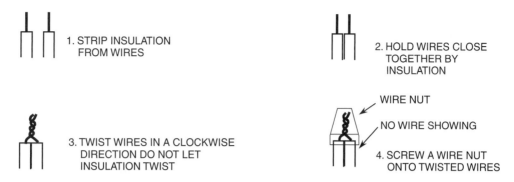

Figure 11-17 **Procedure for splicing solid wires.**

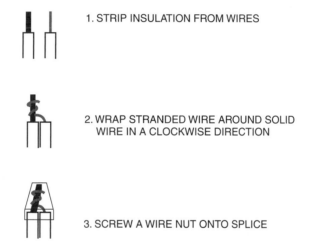

Figure 11-18 **Procedure for splicing solid and stranded wires.**

Push-in Connections (see Figure 11-19)

1. Use the strip gauge on the device to make certain that you strip the correct amount of insulation off the wire.
2. Push the exposed conductor into the opening in the connector. Do not try to force a wire that is too large into the opening.
3. The exposed conductor should go into the opening so that the insulation is right up against the surface of the equipment.

Oil burner wiring, if properly installed, will rarely be a source of service problems. The entire circuit can be checked without removing the wiring from the junction boxes. Once you start to disrupt wires that have been in a junction box for a long time, it may involve replacing some or all of them. When it is necessary for you to do some wiring, you have to avoid building a problem into the system by making a loose splice, jamming too

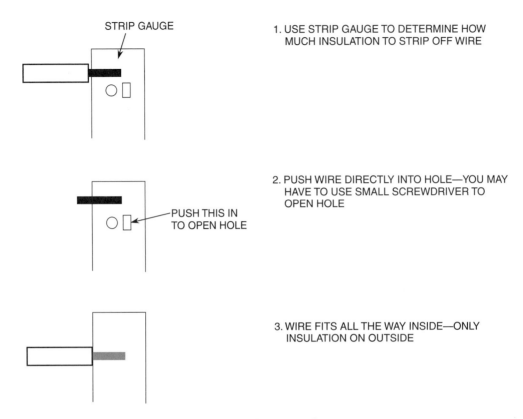

STRIP GAUGE

1. USE STRIP GAUGE TO DETERMINE HOW
   MUCH INSULATION TO STRIP OFF WIRE

2. PUSH WIRE DIRECTLY INTO HOLE—YOU MAY
   HAVE TO USE SMALL SCREWDRIVER TO
   OPEN HOLE

PUSH THIS IN
TO OPEN HOLE

3. WIRE FITS ALL THE WAY INSIDE—ONLY
   INSULATION ON OUTSIDE

Figure 11-19 **Procedure for making push-in connections.**

many wires into a junction box, kinking or making sharp bends in the wire that can create hot spots, or creating a short circuit by grounding an exposed conductor to the metal box or conduit. Once you master the wiring techniques you will find that service calls that are caused by an electrical problem are the easiest to diagnose and repair.

The following definitions will help you in dealing with oil burner electrical wiring and equipment:

1. **Current Electricity**—The movement of electrons through a conductor.
2. **Electron**—A negatively charged particle that orbits the nucleus of an atom.
3. **Direct Current (DC)**—Electrons moving in a straight line without ever changing polarity.
4. **Alternating Current (AC)**—Electrons moving in a wavering line that changes polarity 60 times each second. This is known as 60 Hz (cycle) electricity.
5. **Electromagnetism**—When an electric current moves through a conductor, a magnetic field of force is built up around the conductor. The higher the intensity (amperage) the stronger the magnetism gets.
6. **Single-Phase–120-volt Electrical Service**—One hot wire carrying 120 volts and one neutral wire (see Figure 11-20).

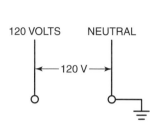

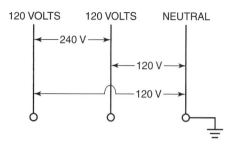

Figure 11-20   Single-phase–120-volt electrical service.

Figure 11-21   Single-phase–240-volt electrical service.

7. **Single-Phase–240-volt Electrical Service**—Two hot wires, each carrying 120 volts and one neutral wire, 240 volts between hot wires (see Figure 11-21).
8. **Three-Phase–240-volt Electrical Service**—Three hot wires, each carrying 120 volts, and one neutral wire, 240 volts between any combination of two hot wires (see Figure 11-22).
9. **Electrical Circuit**—A complete system that includes a source of electrons, a control, a load, and a network of wires that connects the parts.
10. **Series Circuit**—A single path for the electricity to follow. An opening any place in the circuit will shut down the entire circuit.
11. **Parallel Circuit**—Two or more paths for the electricity to follow. The voltage is the same in every branch of a parallel circuit.
12. **Combination Circuit**—Part series and part parallel, these circuits are built to take advantage of the two types of circuits. Oil burner circuits are combination circuits.
13. **Volt**—The unit of measurement of electrical pressure (electromotive force).

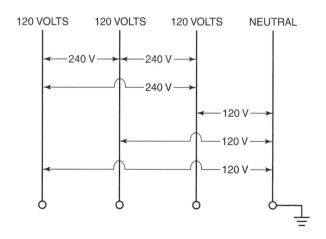

Figure 11-22   Three-phase–240-volt electrical service.

14. **Ohm**—The unit of measurement of electrical resistance.

15. **Ampere**—The unit of measurement of the electrical intensity in a circuit.

16. **Watt**—The unit of measurement of electrical power. Every load requires a specific quantity of watts to operate properly. We pay for electricity by the number of watts consumed within a specific period of time.

17. **Cartridge Fuse**—Protects electrical circuit to prevent fires if too many amperes are sent through the circuit. These fuses fit into a set of copper clips mounted in an electrical panel. The power enters the end of the fuse that is closest to the center of the panel, moves through the fusable link inside the cartridge, and leaves at the end of the fuse closest to the outside of the panel.

18. **Plug Fuse**—Has the same function as the cartridge fuse. The fuse screws into a standard Edison base socket, the same as a standard light bulb. The power enters through the bottom of the fuse, moves through the fusable link, which is clearly visible through the glass top of the fuse, and leaves through the threaded metal portion of the fuse that screws into the socket. A terminal screw is connected to the female threads of the socket. This is known as the load terminal and is where the circuit wiring is connected.

19. **Fustat**—These are upgraded versions of the plug fuse. The socket used for the fustat is a ceramic socket that is color coded so that once the socket is installed only the same size fustat will fit into it. This prevents the homeowner from installing a larger fuse that might lead to a fire or damage the equipment.

20. **Circuit Breaker**—A device that operates as a switch that will turn the circuit off if the ampere rating is exceeded. Circuit breakers are phasing out fuses because they do not have to be replaced, merely reset, if there is an overload or short circuit.

21. **Electrical Wire**—A conductor covered with insulation.

22. **Conductor**—A material that allows electricity to move through it with little resistance. Copper, aluminum, and silver are commonly used conductors.

23. **Insulator**—A material that will not allow electricity to move through it. Plastic, rubber, glass, and porcelain are commonly used insulators.

24. **#14 Wire**—The size of the electric wire used in the 120-volt portions of oil burner wiring. #14 wire is rated for circuits up to 15 amperes.

25. **#18 Wire**—The size of electric wire used in the 24-volt portions of the oil burner wiring.

26. **Romex**—Plastic covered electrical cable that has two or three electrical wires and a bare ground wire inside a heavy plastic armor. 14-2 Romex would have two #14 insulated wires and one #14 bare wire inside the plastic cover. The bare wire is used to ground the electrical components so if there is a short circuit someone touching the parts of the system will not get a shock because the fuse will blow out instantly.

27. **BX Cable**—Two or three electrical wires, each wire wrapped in paper, inside a flexible steel armor. There is a bare aluminum wire inside the armor as well. This wire is a back-up mechanical ground wire, similar in purpose to the bare wire in Romex. The mechanical ground function of BX cable is performed by the external steel armor. 14-3 BX cable is three #14 wires, the aluminum bonding wire, inside the flexible steel armor.

28. **Electric Metal Tubing (EMT)**—Conduit used for surface wiring. EMT can be bent to fit between junction boxes. Wire is then pulled through the EMT into the junction boxes to complete circuits.

29. **Rigid Conduit**—Electrical pipe that is the same weight as water piping. This type of conduit is suitable for underground or outside installation. It must be threaded and is connected to junction boxes by using lock nuts. This type of conduit can be bent to fit between junction boxes, and wires are pulled through the conduit into the junction boxes to complete circuits.

30. **Junction Boxes**—There are a variety of sizes and shapes used in wiring. The main function is to give us a convenient place to mount switches and make splices to complete electrical circuits. The junction boxes are made of steel or plastic.

31. **Mechanical Ground System**—This ensures that in the event of a short circuit in the wiring, the fuse will blow and the people using the equipment will not get a shock. Mechanical ground wiring is usually connected to a green terminal that is mounted to connect the wire directly to the metal case of an appliance or other electrical device. All of the mechanical ground wires must be connected together and run back to the main electrical panel. The panel itself is usually grounded to the water main or grounding rods buried outside the building.

32. **Test Light**—A rubber covered pig-tail socket with a 60-watt light bulb.

33. **Circuit Tester**—A flashlight with test leads that can be used to check for continuity in electrical circuits. These testers must be used with the power turned off because the power for the tester is supplied by the batteries in the flashlight.

34. **Multimeter**—An electrical meter that can check volts, amperes, and ohms. The power must be off when using the ohmmeter. Clamp-on type measures amperes by sensing the strength of the electromagnetic field around a single wire when the power is on.

35. **Splice**—A method of connecting conductors to complete an electrical circuit. The pig-tail splice is the most commonly used in our industry. It is important to connect the conductors tightly to avoid sparking and to make certain that enough surface area of each of the conductors touches so the electrons will have a large surface to flow through. This will avoid hot splices that can lead to serious problems inside a junction box.

36. **Terminal Screw**—A screw attached to an appliance that allows us to connect the wiring to the appliance. Terminal screw connections must be tight; the conductor must wrap around the screw in a clockwise direction so that the wire gets tight under the screw as it is tightened, and the insulation should be as close to the screw as possible, but not under the screw cap.

37. **Toggle Switch**—A single pole, single throw switch used to manually control an oil burner circuit.

38. **Control**—An automatic switch wired into the oil burner circuit. Controls are built to react to changes in one of the following: liquid temperature, air temperature, liquid level, pressure, air movement, and air humidity. By assembling the correct assortment of controls, we can build an automatic heating system circuit that will operate efficiently and safely.

39. **Appliance (Load)**—A device that consumes electrical power. The appliances in an oil burner circuit are the oil burner motor, the ignition transformer, the primary control transformer, and other auxiliary devices.
40. **Motor**—A device that converts electrical energy into mechanical energy.
41. **Step up Transformer**—A device that increases voltage. The ignition transformer increases the voltage from 120 volts to 10,000 volts.
42. **Step down Transformer**—The transformer in the primary control reduces the voltage from 120 volts to 24 volts.
43. **Relay**—An electromagnetic switch that is an essential part of any automatic wiring system.
44. **Solenoid Valve**—An electrically operated valve that can control the movement of liquids or gasses.
45. **Delayed Opening Oil Valve**—A solenoid valve that has a built-in thermister. The thermister will not allow the power to enter the valve coil for approximately three seconds. This allows the oil burner motor to come up to full running speed before the fuel is allowed to enter the combustion chamber, which leads to a smooth light off of the flame. Rapid recycling of the burner will eliminate the delay because the thermister will be heated up and the power will go right through it to the coil.
46. **Time Delay Switch**—A solid-state device that will always give us a time delay when wired in series with a solenoid valve used to control the flow of fuel into the combustion chamber. These switches can be either fixed time or adjustable. Fast recycling of the burner has no effect on the operation of these switches.

To Briefly Summarize

1. Oil burner service technicians must understand basic electrical circuits.
2. Series circuits are used for manual control switches and limit controls because an opening in any part of a series circuit will shut down the entire circuit. Switches are always wired in the hot wire.
3. Parallel circuits are used to wire two or more power-consuming devices in the same circuit because they will get the full voltage required for proper operation.
4. Oil burner wiring is a combination of series and parallel circuits. All wiring must comply with the local building code requirements.

Please Answer These Questions

1. What is the function of the fuse? How does the fuse perform this function?
2. Why are circuit breakers phasing out fuses?
3. What is the mechanical ground system and why is it important?
4. A 120-volt oil burner motor operates at a running amperage of 2.5. How many watts are required for the motor to operate properly?
5. Why are oil burner limit controls wired in series circuits?

6. What is BX cable?
7. What is Romex?
8. How are electrical entrance services identified?
9. What is a combination circuit?
10. Explain the procedure for locating a short circuit in wiring.

ELECTRICAL EQUIPMENT

As an oil burner service technician you will be working on a variety of electrical devices. Each type of device performs a specific function in the system. The diagrams and explanations contained in this chapter will help you to understand how they operate.

SOLENOID VALVES

When an electrical current moves through a wire, a magnetic field of force is formed around the wire. When the wire is wrapped in a coil, the magnetic force is concentrated in the center of the coil. This type of electromagnet can be used to operate a valve that can control the movement of a liquid or a gas. Valves of this type are known as solenoid valves and have these parts (see Figure 12-1):

1. The coil is a continuous length of fine wire insulated with varnish, wound on a spool, and enclosed in a protective covering. The use of a long piece of very thin wire, which has a specific amount of resistance per running foot, makes the coil an appliance that must be connected to a 120-volt circuit. Two lead wires are attached to the coil for this purpose.

2. The valve body has an inlet and outlet port for connection to the piping. These ports have female pipe threads and are clearly marked to ensure that the liquid or gas will travel through the valve in the correct direction. The valve seat is designed to close tightly when the power to the coil is off. Reversing the flow through the valve may cause some leakage through the seat. Care must be taken to keep pipe joint compound out of the valve body. Start the connecting fittings into the ports and then apply the appropriate pipe joint compound before tightening the fittings with a wrench. Avoid using Teflon tape because pieces of the tape could break off and get into the valve seat, causing a leak through the valve. Always use two wrenches when tightening fittings into the valve. One wrench is used to hold the valve body while the other tightens the fittings. This will prevent twisting the valve body, which may cause the valve to leak.

3. The plunger or linkage assembly fits into the valve body. There is a spring that keeps the plunger firmly against the valve seat in the body. The coil fits around the plunger,

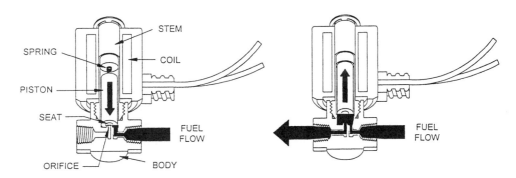

Figure 12-1 **Parts of a solenoid valve.** *Courtesy: R. W. Beckett, Corp.*

or linkage, so that when power is applied to the coil the magnetism created will lift the plunger off the seat and allow the liquid or gas to move through the valve. When the power to the coil is cut off, the spring will drive the plunger against the seat of the valve and close off the flow through the valve.

Delayed-opening solenoid valves are used to control the flow of fuel oil from the pump to the nozzle. These valves allow the burner motor to come up to full speed and the ignition spark to be in position when the oil is released into the combustion chamber. This leads to smooth, smoke-free ignition of the air–oil mixture. Another benefit is the clean cut off at the end of the burner cycle due to the snap action closing of the valve when the power is shut off.

The delay action of the valve can be accomplished in one of two ways. The first type has a solid state electronic device, a thermister, permanently wired in series with the valve coil. This small unit fits inside the coil cap. The current to the coil is slowed down by the thermister for approximately three seconds. Once the thermister heats up, it allows the current to continue and the valve opens. Fast recycling of the burner will eliminate the delay because the thermister must cool off to be effective. The second method employs an external device that is wired in series with the coil (see Figure 12-2). These are not affected

Figure 12-2 **Time delay switch—Honeywell ST 70.** *Courtesy: Honeywell.*

by rapid recycling and will always deliver the desired delay timing. There are fixed timing and adjustable timing devices available. It is important to remember that these units are not appliances and must be wired in series, usually in the hot wire, with the coil. Connecting a hot and neutral wire to any of these time delay units will burn them out.

Another use of solenoid valves is as a boiler feed valve in a steam heating system. The valves designed for this purpose are much larger than the oil valves but have basically the same parts and operate in exactly the same way (see Figure 12-3). It is essential that the valve manufacturer's installation instructions be followed carefully. Of particular importance is the direction of flow through the valve. Some general suggestions for installing water feed solenoid valves are:

1. You must check the water in the area before suggesting the installation of a solenoid water feed valve. Hard water or water with impurities will adversely affect the operation of the valve, which will lead to service problems.
2. Connect the valve to the cold water piping as hot water can damage the valve seat.
3. Install a manual valve before the solenoid, and another manual valve to act as a bypass, which will put water into the boiler if the solenoid valve fails.
4. Install the solenoid and accompanying valves above the water level of the boiler.
5. Install unions on the piping just before and just after the solenoid valve. This will make checking the operation and replacing the valve easy (see Figure 12-4).

Figure 12-3 #101 electric water feed valve. Hexagonal hub at piping ports indicates that the feeder is equipped with a replaceable cartridge valve. *Courtesy: McDonnell Miller*

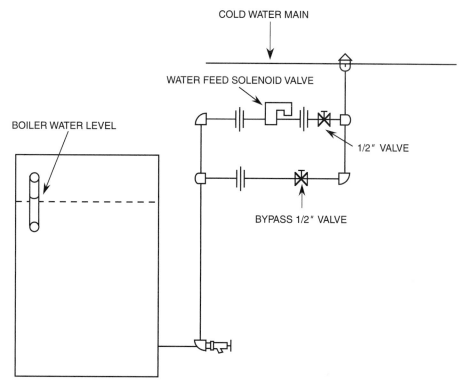

Figure 12-4   Piping hook-up for the McDonnell Miller #101.

RELAYS

Another use of electromagnets is to operate electrical switches called relays. Relays are integral parts of all automatic control systems. Relays all have these basic parts (see Figure 12-5):

1. The relay coil is made to operate at a specific voltage. Each oil burner primary control has at least one relay. These relay coils are built to become magnets when the 24 volts supplied by the transformer in the primary controls are applied to the coil. This arrangement allows the use of a small, sensitive control to operate the oil burner. There are other relay coils that require 120 volts to operate. There are even relay coils made to operate at 240 volts. The coil, which is a part of the entire unit, must be able to deliver enough magnetic force to close the relay contacts. The size of the coil and the operating voltage will depend on the designed application for a particular relay.
2. The magnet fits inside of the coil. This is a temporary magnet made of a special alloy that will lose its magnetism as soon as the power to the coil is shut off. This ensures that the relay will close when the power to the coil is on, and open when the power to the coil is off.

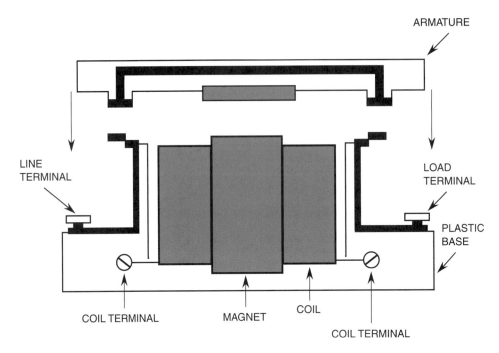

Figure 12-5   **Single pole, single throw relay.**

3. The armature has a metal part that is attracted to the relay magnet. This is the part of the relay that actually moves to close the contacts. A spring is attached to the armature to pull it away from the magnet when the power to the coil is shut off.
4. The line contacts of the relay are for the incoming power to the relay switch. They are connected to terminal screws attached to the plastic base of the relay. A relay can have a number of line contacts, depending on the particular application.
5. The load contacts are mounted on the relay armature and are attached to terminal screws in the plastic base of the relay. These contacts line up with the line contacts so that when the relay closes, electricity will flow from the line contacts through the load contacts to the appliances wired to the relay (see Figure 12-6).

There are several switching arrangements utilized in the design of relays. The simplest is the single throw, which means that the contacts close when the coil is energized and open when the power to the coil shuts off. A single pole, single throw relay will complete a single circuit when the coil is energized. A double pole, single throw relay will complete two circuits when the coil is energized (see Figure 12-7). A three pole, single throw relay will complete three circuits when the coil is energized and so on depending on the number of line and load contact sets built into the relay. Double throw relays complete one set of circuits when the coil is energized and another, completely separated set, when the coil is deenergized (see Figure 12-8). The line wiring to these relays is connected to the armature with the number of circuits available determined by the number of sets of line and load contacts.

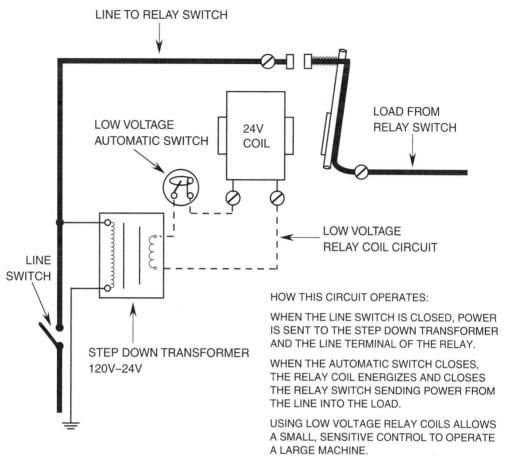

LINE TO RELAY SWITCH

LOW VOLTAGE
AUTOMATIC SWITCH

24V
COIL

LOAD FROM
RELAY SWITCH

LOW VOLTAGE
RELAY COIL CIRCUIT

LINE
SWITCH

STEP DOWN TRANSFORMER
120V–24V

HOW THIS CIRCUIT OPERATES:

WHEN THE LINE SWITCH IS CLOSED, POWER
IS SENT TO THE STEP DOWN TRANSFORMER
AND THE LINE TERMINAL OF THE RELAY.

WHEN THE AUTOMATIC SWITCH CLOSES,
THE RELAY COIL ENERGIZES AND CLOSES
THE RELAY SWITCH SENDING POWER FROM
THE LINE INTO THE LOAD.

USING LOW VOLTAGE RELAY COILS ALLOWS
A SMALL, SENSITIVE CONTROL TO OPERATE
A LARGE MACHINE.

Figure 12-6 **Line-load and relay coil circuits. Operation of a simple control circuit with a
low voltage relay coil.**

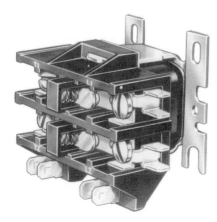

Figure 12-7 **Double pole, single throw relay—
Honeywell R4242.** *Courtesy: Honeywell.*

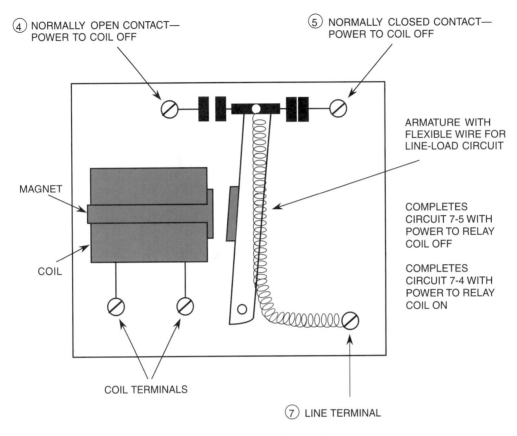

④ NORMALLY OPEN CONTACT—
POWER TO COIL OFF

⑤ NORMALLY CLOSED CONTACT—
POWER TO COIL OFF

ARMATURE WITH
FLEXIBLE WIRE FOR
LINE-LOAD CIRCUIT

MAGNET

COIL

COMPLETES
CIRCUIT 7-5 WITH
POWER TO RELAY
COIL OFF

COMPLETES
CIRCUIT 7-4 WITH
POWER TO RELAY
COIL ON

COIL TERMINALS

⑦ LINE TERMINAL

Figure 12-8 **Operation of a single pole, double throw relay.**

There are separate circuits connected to the relay, the coil circuit and the line-load circuit (see Figure 12-6). The coil circuit allows the use of small controls and switches because it takes very little electricity to energize a relay coil. This means that the line-load circuit is where the heavier wiring and switches are found. The relay contacts are built to withstand the high amperage draw created by the starting surge of a motor. This makes the relay a perfect switch to operate a motor in an automatic system. Motor starter relays have built-in overload protection for the motor.

TRANSFORMERS

Another use of electromagnetism is current induction in transformers. Transformers all have these parts (see Figure 12-9):

1. A primary coil that is connected to a 120-volt circuit. This coil is made to fit around an iron core.

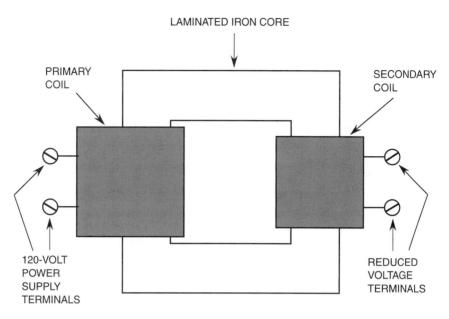

Figure 12-9 **The parts of a step down transformer.**

2. A secondary coil that is also built to fit around the iron core. If the secondary coil is smaller than the primary coil, the voltage induced in the transformer will be less than the 120 volts fed into the primary. This is true in the step down transformer built into the primary control to supply the power for the thermostat circuit. If the secondary coil is larger than the primary coil, the voltage in the secondary coil will be higher than 120 volts. This is true in the ignition transformer that has two large secondary coils wrapped around the core.

3. The core is made up of a series of soft steel plates. These are known as laminated cores and are built to accept the primary and secondary coils.

When power is applied to the primary coil, it becomes an electromagnet. The magnetism from the coils also makes the soft steel core into a magnet. Because of alternating current flowing through the primary coil, the polarity of the magnetic field inside the coil switches 60 times a second. This causes the polarity of the soft steel core magnet to switch at the same frequency. The switching polarity inside the secondary coil forces the electrons—negatively charged particles in the coil—to move toward the positive pole and away from the negative pole of the magnet. Obviously, the greater the number of electrons involved in this frantic switching, the higher the voltage will become.

MOTORS

The ultimate use of electromagnetism is to drive electric motors. An electric motor is a device that converts electrical energy into mechanical energy. Think what life would be without all of the appliances and tools we use that have electric motors built into them. It would take a series of text books to thoroughly cover the subject of electric motors. All electric motors have similar parts. The differences in motors have to do with their internal wiring. We will be discussing the types of motors used in the vast majority of gun type oil burners—the split-phase motor and the permanent split-capacitor motor.

All motors can be broken up into four major assemblies: the front end bell, the stator, the rear end bell, and the rotor and shaft assembly (see Figure 12-10).

The Split-Phase Motor

The split-phase motors are assembled as follows (see Figure 12-11):

1. The front end bell that has a carefully machined rabbet (a precisely machined ridge) fits into the burner housing perfectly; a two-hole flange that lines up with the mounting tappings on the burner housing; the front bearing and its lubrication channel; and an opening for the shaft that extends approximately 4″ from the face of the end bell.
2. The stator, that has a laminated iron core with slots in it for the motor windings. The windings are coils of wire with varnish insulation that are set into the core in an exact pattern. There are two sets of windings in the core. The running winding is made of coils that supply the magnetic force to keep the motor running over long periods of time. The starting winding is made of lightweight coils that supply the additional magnetic force to start the motor. These coils can only be in the circuit for a few seconds at a time or they will burn out. The stator case has a rabbet edge that fits into the front end bell.

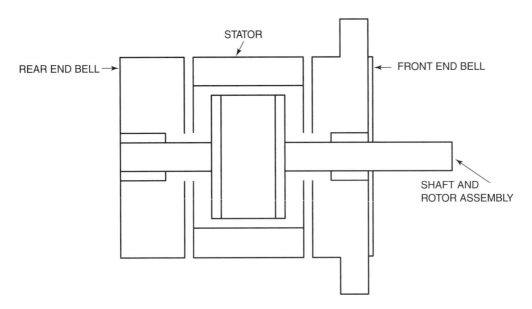

Figure 12-10 **The four main assemblies of an oil burner motor.**

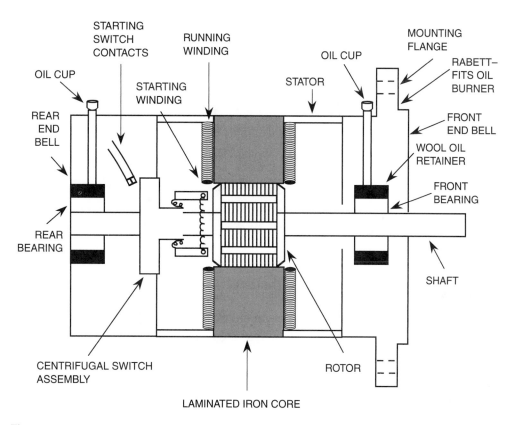

Figure 12-11 **The parts of a split-phase oil burner motor.**

3. The rear end bell fits precisely into the stator case. The rear bearing and its lubrication channel are set into the rear end bell. The stationary portion of the starting switch can be mounted in either the front or rear end bell. The wiring connections to the motor and the thermal overload device are all part of this assembly. There are four holes drilled in the end bell for the long through bolts that slide through holes in the stator into holes in the front end bell. Once the through bolts are tightened the motor is completely assembled.

4. The motor shaft assembly fits precisely in the center of the assembled motor (see Figure 12-12). The shaft end, which extends from the front end bell, is 1/2″ in diameter and has a flat surface machined on it to accept the fan and oil pump coupling set screws. Just inside the front end bell is the bearing surface that fits into the front bearing. This surface is perfectly machined to a mirror smooth finish, which allows the shaft to turn in the lubricated bearing without wearing it out. The rotor is mounted on the shaft so that it lines up perfectly with the stator core. The rotor is made of soft iron plates pressed tightly together and machined into a perfect circle. Imbedded in the rotor are copper or aluminum rods that are attached to similar material rings on the ends of the rotor. As the magnetic fields created by the stator sweep around the rotor,

Figure 12-12 **Split-phase motor with front end bell removed. The centrifugal switch is mounted on the shaft. The switch contacts and thermal overload switch are mounted on the front end bell.** *Courtesy: Carlin Combustion Technology, Inc.*

the electrons in the rods begin to move and create a current that sets up a magnetic field around the rotor. The magnetism from the stator and the opposing magnetism of the rotor drive the rotor into its spinning motion. The weight of the rotor and its spinning motion deliver the mechanical force that can drive a machine. The spacing between the rotor and the stator core is very close, approximately 1/32". The rotor may have a set of blades attached to it to act as a cooling fan for the motor. Attached to one end of the rotor, to line up with the stationary starting switch, is the starting switch clutch assembly. This mechanism is a set of weights and springs that react to centrifugal force when the rotor starts to turn. There is a large plastic ring set into the assembly that forces closed the starting switch contacts in the end bell when the motor is standing still. When the motor starts, centrifugal force moves the plastic ring away from the switch, allowing the contacts to open. You can hear the click of the switch opening when the motor starts and you can also hear it click back to the starting position when the motor stops. The starting switch is wired in series with the starting winding of the motor so that once the motor starts to turn fast enough to move the centrifugal clutch, the starting winding magnetism is no longer necessary and the starting winding can be taken out of the circuit. The end of the shaft has the rear bearing surface that is machined to a mirror finish and fits precisely into the rear bearing.

Sequence of Operation of Split-Phase Motors

The split-phase motor derives its name from the starting sequence. When the power to the motor is off, the shaft is standing still in the bearings. The centrifugal clutch presses the starting switch contacts closed. The motor is ready to run. When the power is turned on, voltage flows to both the starting and running windings. Because the starting winding is made of smaller wire than the running winding it has higher resistance, which slows the current flow in the starting winding. This slow-down of the current flow splits the phase, creating two overlapping magnetic fields in the stator. These magnetic fields induce a current in the rotor, which then establishes its own magnetic field that is out of step with the magnetic fields created by the starting and running windings. The separated magnetic fields drive the rotor into its spinning motion. When the rotor is turning at approximately 75% of its normal running speed, the centrifugal clutch opens (you can hear the click) and shuts off the power to the starting winding. The magnetism created by the running winding will continue to drive the rotor, which will come up to its normal running speed. When the power to the motor shuts off, all of the magnetism stops and the motor winds down to a stop. You can hear the click as the centrifugal clutch returns the starting switch to the closed position so that the motor is once again ready to start.

Permanent Split-Capacitor Motors

Permanent Split-Capacitor Motors Are Assembled as Follows (See Figure 12-13):

1. The front end bell has a carefully machined rabbet that fits into the burner chassis perfectly, a two hole flange that lines up with mounting tappings on the chassis, the front ball bearing, and an opening that the shaft extends through. The face of the end bell is flat, which allows the fan to be mounted on the shaft very close to the face. This eliminates loss of air velocity when the burner is running.
2. The stator, which fits securely into the front end bell, has the laminated iron core and both the main and auxiliary windings. The windings are made of the same size wire

Figure 12-13   **Cutaway view of a split-capacitor motor.** *Courtesy: R. W. Beckett Corp.*

and are offset from each other to provide the magnetic forces necessary to drive the rotor. The capacitor is enclosed in a container attached to the outside of the stator.

3. The rear end bell contains the rear ball bearing and fits into the stator. When the shaft assembly is in place, the through bolts are installed from the rear end bell into the front end bell where they are tightened, thus holding the motor securely together.

4. The rotor and shaft assembly fit perfectly inside the motor. The rotor, a heavy laminated cylinder that has aluminum rings on each end with connecting aluminum rods embedded in the laminations, lines up with the laminated core of the stator. The rear shaft end fits securely in the rear bearing and the front bearing surface fits securely in the front bearing with the shaft extending 4″ from the face of the motor.

Sequence of Operation of Permanent Split-Capacitor Motors—PSC Motors

As you can see in the simplified wiring diagrams in Figure 12-14, the PSC motors are not very different from the split-phase motors they will eventually replace. The capacitor takes the place of the starting switch and both sets of windings are made of the same size wire in the PSC motors. The sequence of operation of both motors is very similar. In the PSC motors, the power is applied to the main winding and the capacitor at the same time. The capacitor delays the power for a fraction of a second and then allows it to pass through to the auxiliary winding. This fraction of a second lag establishes two separate, but equal, magnetic fields inside the motor. This magnetism induces a current in the rotor, which then generates its own magnetism and reacts to all of this intense magnetism by spinning, and the motor is running. All of this, a sort of chain reaction, takes place in an instant.

There are several advantages the PSC motors have over the split-phase type. First is the elimination of the starting switch, which has been a source of motor breakdowns in the past. Second is the use of ball bearings that can be permanently lubricated and do

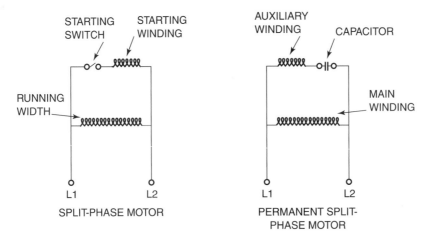

Figure 12-14 **Simplified wiring diagrams for split-phase and permanent split-capacitor motors.**

not have the problem of end play. Allowing the shaft to move in and out of the motor even slightly can affect the operation of the burner. They are also more economical to operate and have sufficient starting torque to operate the oil burner.

CAPACITORS

The ability of a motor to start a machine turning from a standstill is known as the starting torque. Split-phase motors have a starting torque of 150 to 200% of the full load torque. This is important because it is harder to start the machine than it is to keep it running. It should be noted that the starting current draw will be 6 to 8 times the running or full load amperage as well. When a large oil burner with a large fan starts, we need a higher starting torque than can be delivered by a split-phase motor. The higher starting torque can be delivered by a capacitor start split-phase motor. The capacitor is a device that is wired in series with the starting winding. The starting winding in this type of motor is made of the same size wire as the running winding. The splitting of the phase is accomplished by the capacitor that creates a wider split than the difference in resistance between the two sets of windings can. The wider split has the effect of increasing the starting torque 300 to 350% of the full load torque. The capacitor is housed in a separate container attached to the outside of the motor and is wired in series with the starting switch.

BEARINGS

All motors have bearings, as do all machines that have shafts that turn. The functions of the motor bearings are to support the shaft so that it remains lined up perfectly in the center of the motor, and to allow the shaft to spin freely with as little friction as possible. There are two types of bearings used in oil burner motors, the sleeve bearing and the ball bearing.

Sleeve bearings are made of bronze or a similar soft metal. These bearings are lubricated with #10 motor oil, which is held around the bearing by a wick made of wool felt. The bearing has a hole in its top to allow oil to seep into the space between the bearing and the highly polished surface of the motor shaft. The space is very tiny, perhaps .002″, and the bearing may have a set of channels for oil machined into its inner surface. Proper lubrication of these bearings is essential to the life of the motor. A few drops of #10 oil once a month will keep the bearings lubricated. Over-oiling the motor can cause damage and can lead to having to replace the motor.

Ball bearings are much stronger than the sleeve bearings. The inner ring of the bearing is pressed onto the motor shaft. A series of hardened steel balls fits into the other side of the inner ring. An outer ring completes the bearing assembly and is the part that is pressed into the motor end bell. The bearings are permanently lubricated with grease and can give many years of uninterrupted service.

The Motor Nameplate

The motor nameplate, usually attached to the stator of the motor, contains the following information:

1. Manufacturer's name
2. Model number
3. Serial number
4. Size of the motor in horsepower (HP)
5. Required voltage
6. Running amperes
7. Speed in rpm
8. Frame number
9. Temperature rise
10. Safety factor

This information is important when a motor has to be checked out or replaced. Of particular interest when replacing a motor are the speed, frame number, size, and direction of rotation. Some oil burners are built to operate at 3,450 rpm. Installing a 1,725-rpm motor will obviously result in improper operation of the burner. The frame number is actually a physical description of the motor. Most oil burners utilize #48 frame motors. The letter designation after the frame number will indicate the location of the mounting flange holes and where the power leads or terminals are located. It is always advantageous to have an exact replacement but you can certainly adapt another motor with a #48 frame when you have to. The size of the motor should not be changed. Installing a larger (in HP) motor will not create problems, but installing a smaller motor can lead to service calls due to the motor tripping its overload switch. The direction of rotation should be checked before the replacement motor is installed on the burner. Connect the power leads to the motor and run it for a few seconds to see if it is turning in the correct direction. The rotation of many motors can be changed by switching two leads at the starting switch. When a motor is of this reversible type, directions for changing the rotation will be on a small plate that covers the starting switch terminals.

Servicing Motors

Checking a motor requires a combination volt and amp meter. The voltage must be checked first. If the voltage is correct then the amperage can be checked. The proper running amperage is listed on the motor nameplate. If the running amperage exceeds the listed amperage, the overload switch may trip out and shut the motor off. High running amperage may be caused by tight or unlubricated bearings, a fuel pump that is hard to turn due to problems with the gears, or a motor that is too small for the burner.

Low voltage will cause the amperage to increase. This is true only with motors, which always try to run, and need the proper amount of watts to operate. The watts required are not listed on the nameplate; you have to multiply the amps by the volts to determine the required watts. For example, a motor nameplate states that the volts = 120 and the running amps = 2.5. The required watts are $120 \times 2.5 = 300$ watts. This motor will always need 300 watts regardless of the voltage. Reducing the voltage to 100 volts would automatically increase the required amperage to 3 amps. The safety factor on the nameplate will indicate how high the amperage can be before the overload trips. The temperature rise will indicate how hot the motor will feel when it is running. The temperature rise, indicated on the nameplate, does not take into consideration the ambient temperature. If the motor is running in a hot boiler room, its normal temperature rise could damage the motor. It is important to have a good flow of air around any operating motor so that it does not overheat.

As an oil burner service technician, you will not be expected to repair motors in the field. However, it may be possible to replace a defective capacitor and get a motor running. To check the capacitor, disconnect the power to the motor and take it out of its protective cover. You will see that there are two wires connected to terminals (usually stake type terminals) on one end of the capacitor. Be extremely careful and be certain the power to the motor is off! The capacitor can store an electrical charge, so it is necessary to discharge the unit before you can test it. Remove the wires from the capacitor terminals using an insulated pair of pliers. Hold the blade of a well-insulated, dry screwdriver across the terminals. You will get a rather large spark as the capacitor discharges. If there is no spark, the capacitor is defective. Check the rating, shown in microfarads (MFD) on the side of the capacitor and make an exact replacement to get the motor back in operation.

Another method of testing capacitors is with an ohmmeter. For this test you will need a sensitive analog "needle" type tester, not a digital unit. Be certain to have the power off and the wires to the capacitor disconnected before discharging the capacitor as explained above. Hold one lead from the tester to one of the terminals and then touch the other terminal with the other lead. If the capacitor is okay, the meter needle will momentarily jump off the infinite stop and then quickly return to infinity. This happens quickly because the capacitor will become charged from the battery in the ohmmeter and then resist any change. If the meter reads 0 ohms the capacitor has short-circuited and will have to be replaced. If the meter never moves off the infinity stop, the capacitor has an open circuit and will have to be replaced. Remember to be safe and discharge the capacitor after the test. Even the small amount of current from the meter battery can build up in the capacitor and give you a nasty sting if you are not careful. If the capacitor checks out as being good and there is the proper voltage at the motor and it still doesn't run, you will have to replace it.

Always try to make an exact replacement of a defective motor. Your service van inventory should include several motors that can be used on the burners you are expected to repair. Most of the gun type burners use a #48 frame motor, which makes it possible to install the wrong motor for a particular burner. Read the nameplates carefully and pay particular attention to the horsepower rating. Do not replace a 1/5 HP motor with a 1/7 HP motor. That will not work, even though the smaller motor fits on the burner chassis, so don't waste your time.

Some other things to be aware of are the position of the fan and the rotation of the motor. The fan must be as close to the face of the motor as possible, particularly if the burner is one of the forced draft units. This can be accomplished by placing a flat piece of metal between the fan and the face of the motor as a gauge. The Beckett T Gauge is ideal for this purpose. The rotation required must be checked before you mount the new motor on the burner chassis. Most of the split phase motors can have their rotation changed by switching two wires on the starting switch. These wire terminals are under a plate that has instructions for changing rotation printed on it. Please install the wiring to the motor correctly so that the burner looks as good as new when you leave it.

ZONE VALVES

The zone valves used in hydronic heating systems are of the heat-actuated or motor-driven type. Both types of valves feature slow opening and closing, which results in smooth flow of the water through the system. The heat-actuated type has a thermostatic coil, attached to the valve mechanism. When power is applied to this coil it expands and opens the valve. When the power is turned off, the coil contracts to its original position and closes the valve. Figure 12-15 shows a heat-actuated zone control valve. The motor-driven valves have a motor attached to the valve linkage. This motor does not make a complete revolution. It turns approximately 180 degrees to open the valve. The linkage is spring loaded to close the valve when the power to it is shut off.

Zone valves regardless of type have these two features. There is a method of opening the valve manually in the event that the power supply or controlling thermostat should malfunction. The second feature is an "end switch." This switch closes when the valve opens. The end switch can be wired to turn on either the circulator or oil burner, or both, when the zone thermostat calls for heat.

Figure 12-15   **Heat-actuated zone valve—Honeywell V8043.** *Courtesy: Honeywell.*

The power supply for zone valves is a step down transformer. Please use the transformer designed for the valve as explained in the manufacturer's literature. It is possible to use a single transformer for several valves but please check the information sent with the valves and transformers to avoid building a service problem into a system.

ELECTRICAL TESTING EQUIPMENT

In order to determine if an electrical appliance is defective, we must first determine if it is receiving its proper supply of electrical power. To do this we need to use one of the following devices:

1. The test lamp is a simple pig-tail socket with a 60-watt bulb in it. To use this tester to test for power, hold the white wire against a grounded surface (a water pipe, the junction box or conduit) and then touch the black wire to a terminal screw or exposed conductor (see Figure 12-16). If the light comes on, you can be sure that there is power at that point. If you are testing for power at a switch, you must open the switch and then touch each of the terminal screws on the switch. The light will come on when you touch the feed wire to the switch. Close the switch and see if there is power on

Figure 12-16 **Using a test light to check for power at a control. The white lead is grounded to the control case. The light comes on indicating the power is on.**

the other terminal. If there is power, the switch is good. If there is no power, the switch is defective. If you are testing the power at an appliance, you must touch the white wire to the neutral wire and the black wire to the hot wire. If the light comes on, you can be certain that there is power to the appliance and if it is not operating, it is defective. Of course if the light does not come on, you will have to find out why because that is why the appliance is not operating.

2. The voltmeter (see Figure 12-17) is used when you have to determine if the problem with a particular load or appliance, a motor or transformer, is due to a low voltage condition in the building. The voltmeter has two leads extending out of it. Some meters require that the polarity be maintained in a specific way. For that purpose, the test leads will be of different colors and the instructions will tell you which lead to use for the neutral connection and which one to use for the hot connection. The meter is used exactly like a test light. Some meters have more than one voltage scale so it is necessary to set the meter to the scale that is closest to the voltage of the system you are working on.

3. An ammeter is used to check the ampere draw of a load. This is particularly important when motors have to be checked. There are two types of ammeters, the clamp-on and the series. The clamp-on ammeter (see Figure 12-18) has a set of jaws that open so that they can be placed around the hot wire going to the load. These meters

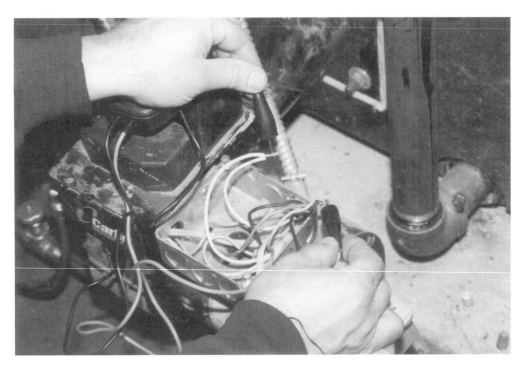

Figure 12-17 **Using a voltmeter to check the voltage at the motor wiring in the junction box on a Carlin CRD 100 oil burner.**

Figure 12-18 **Using a clamp-on ammeter to check the amperes drawn by an oil burner motor. The clamp is placed around the hot wire.**

have several scales and you have to set the meter to the scale that is closest to the ampere ratings you are working with. When the power to the load is turned on, the magnetic field of force generated by the amperes will deflect the indicator needle in the meter. The higher the ampere flow the stronger the magnetic field becomes and this can easily be read on the meter scale. For low ampere applications, the series meters can be used by wiring the hot wire in series with the meter and then on to the load.

4. The ohmmeter is used to measure the resistance in a circuit and to determine if there is an opening in a circuit or to find out if there is a short circuit (see Figure 12-19). It is essential that the power to the circuit being tested is disconnected. Ohmmeters have a battery to supply their power, and connecting an ohmmeter to a circuit that has power flowing through it will destroy the meter. To determine the resistance in a load, connect the ohmmeter to the two leads attached to the load. The meter will read the ohms of that particular load. To determine the resistance in a circuit, connect the ohmmeter to the hot and neutral wires going to the circuit after isolating the circuit from the power supply. A reading of 0 ohms would indicate a short circuit. A reading of infinity would indicate an open circuit. Any reading between these two extremes would be the actual resistance in the circuit.

5. The flashlight circuit tester (see Figure 12-20) can be used to find grounded wires and open circuits. These testers have batteries that supply their power and must be

Figure 12-19 Using an ohmmeter to check for grounded wire. The power must be off. A reading of infinity means there is no ground. A reading of 0 ohms indicates a grounded wire.

Figure 12-20 A flashlight circuit tester checking for a grounded wire. No light means no ground. If the light comes on, the wire is grounded.

used with the power supply disconnected or they will be damaged. To find a grounded wire, attach one lead from the tester to the side of a junction box. Then touch the other wires in the box. When the light comes on, that is the grounded wire because there is a completed circuit from the flashlight through the wire to the conduit and back to the junction box. To test continuity in a long run of wire, you can disconnect and ground one end and then test the other end. If the light comes on, the wire is continuous. If the light does not come on, the wire is broken.

To Briefly Summarize

1. Servicing electrical equipment is an integral part of oil burner service work.
2. Always check the incoming power to an electrical device before deciding that it is defective. Use a test lamp of the appropriate voltage or a multimeter that can test both volts and amps.
3. Try to make exact replacements if possible. If you do not have an exact replacement, be certain to match the volts, size, and the mounting arrangement. Replacement parts must fit and the completed job should look as good as new.
4. Always follow the manufacturer's instructions when installing or replacing electrical equipment.
5. All wiring must be done in compliance with the local electrical code.

Please Answer These Questions

1. Explain the operation of a solenoid valve.
2. How are solenoid valves used in automatic heating systems?
3. Identify the parts of a relay.
4. Why are low voltage relay coils important in automatic heating system control circuits?
5. Explain how a transformer can change the voltage.
6. What are the functions of motor bearings?
7. Explain the operation of the starting switch in a split-phase motor.
8. How is a capacitor used in a motor and when is it necessary?
9. Explain why reduced voltage may cause the motor overload to trip.
10. What is the function of the end switch on a zone control valve?

OIL BURNER CONTROLS

Automatic heating systems depend on controls to keep them operating safely and efficiently. A control is actually a switch that is operated by some kind of mechanical device. The mechanism used depends on exactly what we want the switch to monitor. We can set up controls to monitor liquid level, liquid temperature, steam pressure, air pressure, air temperature, air movement, and the oil burner flame. By installing the right combination of controls for a specific system, and wiring them in a circuit that will operate to our specifications, we get the desired result; a safe, efficient system.

Oil burner controls can be divided into three categories:

1. Limit controls that are designed to keep the system operating within safe limits.
2. Operating controls that are designed to turn the system on when there is a demand for heat or hot water.
3. Primary controls that are designed to specifically monitor the operation of the oil burner itself.

There is some crossover between the limit and operating controls, depending on how they are wired in the circuit.

WATER LEVEL CONTROLS

Boilers for steam or hot water heating systems depend on the water inside them to absorb the heat released by the oil burner flame. If there is not enough water in the boiler, the oil burner flame will overheat the boiler and either crack it or open one of the welded seams. In either event, the boiler may have to be replaced, which is extremely expensive. The limit control that monitors the water level is called the low water cutoff. There are two types of low water cutoffs now in use: the float operated and the electric probe type.

Boiler manufacturers may provide a 2-1/2″ tapping at the minimum safe water level so that a low water cutoff can be screwed directly into the boiler (see Figure 13-1). This type of control has a float that is actually in the boiler water. As the water level in the boiler drops, the float sinks lower in the control. The float is attached to a switch on the outside of the boiler by a linkage device. There is a bellows-type seal used to keep water from leak-

Figure 13-1   **Low water cutoff designed to be installed in a 2-1/2″ boiler tapping.** *Courtesy: McDonnell Miller.*

ing into the switch while still allowing the float to rise and fall freely with the water level. The switch has four terminals and three operating positions (see Figure 13-2). When the float is up, indicating that there is enough water in the boiler, there is a complete circuit between the two top terminals, #1 and #2. These terminals are used for the limit control circuit of the oil burner. As the water level drops, the second circuit is closed, while the first one remains closed as well between terminals #3 and #4. This portion of the switch can be wired to an electrically operated water feed valve that adds water to the boiler every time the float drops low enough to activate this circuit. If the water level continues to drop, the top switch between terminals #1 and #2 will open and shut off the oil burner, while the bottom switch terminals remain closed to continue adding water to the boiler.

Another version of the float-operated low water cutoff mounts at the gauge glass tappings of the boiler (see Figure 13-3). The internal mechanism and the switch are the same as the screw-in type. Both units include a large area opening blow down valve at the bottom of their water chambers (see Figure 13-4). As you can see, there is not a lot of room between the float and the walls of the chamber. It is absolutely essential that the water chamber be kept free of sediment build-up which could cause the float to stick. The owner of the building must be instructed to maintain this control by draining water out of the blow down valve on a regular basis. Service technicians can check the operation of the control by opening the blow down valve while the burner is operating. Once the water starts coming out of the control, check to see if the water feed valve opens and then if the burner shuts off. If the burner does not shut off when the water is drained, you can be certain the float is stuck and the control will have to be taken apart and cleaned.

The low water cutoff can also be part of a mechanical water feeder (see Figure 13-5). These float operated units are connected to the boiler at the gauge glass tappings

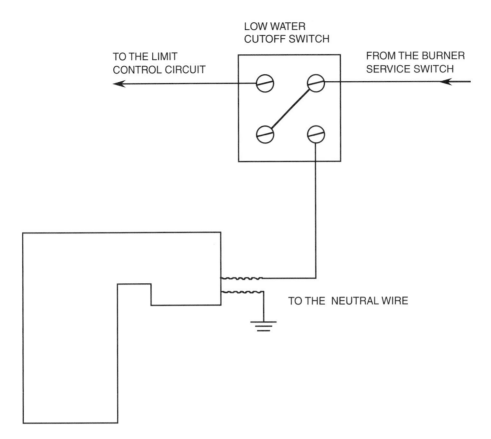

Figure 13-2   **Wiring the low water cutoff to operate an electric boiler water feed valve.**

by the use of specially designed fittings for quick hook-up. These units have a much larger float and double linkage that will open the water valve, designed specifically for this control, and operate the switch that is wired in the oil burner limit control circuit. The large float is necessary because the mechanical force that is applied to close the valve is equal to the weight of the water displaced by the float. The small float used in the other low water cutoffs would not be able to supply the necessary force. This type of control is quite large and easy to spot. It usually has the same type of blow down valve at the bottom of the float chamber for easy flushing. The water supplied to the valve must be cold. Hot water will damage the stainless steel seat of the valve and cause it to leak. If the valve leaks, even a slow drip, the boiler will eventually flood. These combination controls are designed specifically for low pressure steam heating systems. It is essential to follow the manufacturer's instructions when connecting the water piping to the feed valve (see Figure 13-6).

The electrical probe type of low water cutoff is becoming increasingly popular for steam as well as a back-up safety device for hot water heating systems (see Figure 13-7). The probe is installed in a 3/4" tapping at the safe minimum water level of the boiler. The

Figure 13-3 Low water cutoff designed to be mounted at the boiler gauge glass tappings, has the 14-B ball-type blow down valve. *Courtesy: McDonnell Miller.*

Figure 13-4 14-B ball-type blow down valve for McDonnell Miller low water cutoffs. Full port opening allows easy flushing of scale and rust. *Courtesy: McDonnell Miller.*

Figure 13-5 **Combination mechanical boiler water feeder and low water cutoff is mounted at the boiler gauge glass tappings.** *Courtesy: McDonnell Miller.*

probe itself is similar to an electrode. It is insulated by a high-temperature resistant porcelain insulator. The metal portion of the probe, which extends into the boiler water, may be attached to a strip of metal that increases its surface area. The metal probe is in contact with the boiler water that acts as the ground for the relay coil and operates the switch inside the control. The power for the relay coil circuit is supplied by a small step up transformer built into the control. When the probe is in contact with the boiler water, the relay coil is energized, which keeps the limit control circuit switch in the closed position, allowing the burner to operate. If the water level drops below the probe, the relay coil deenergizes, opening the relay and shutting the burner off. As you may know, when a steam boiler is operating, the water can be bouncing up and down inside the boiler. To eliminate the problem of short cycling the burner because of the surging of the boiler water and the loss of contact between the water and the probe, a time delay device is included in the control.

Electronic water feed valves have been developed to operate as a team with the probe type low water cutoffs (see Figure 13-8). It is necessary to install these valves in matched sets or to use a valve that can be adjusted to the timing circuit of the low water cutoff on the job. In any case, it is essential to follow precisely the manufacturer's installation instructions to ensure proper operation. These instructions are always included in the control or valve packaging. Manufacturers spend a great deal of money preparing these information sheets and it is absolutely essential for you to read and follow these instructions.

How a Feeder Cutoff Combination Works

Normal Operation
This drawing shows a typical installation of a McDonnell Combination Boiler Water Feeder and Low Water Cutoff (No. 47–2) on an average steam boiler. Notice that the control is installed well below the normal operating boiler water level. This allows ample space in the boiler for normal evaporation and the return of condensate without the necessity of the Feeder adding any water. The automatic burner is ON, and the boiler is functioning in the usual manner. The Feeder is merely "standing by."

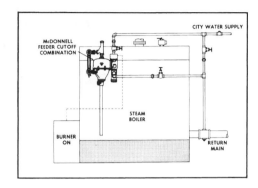

Feeder Operation
If the boiler water line should drop below its normal operating range, the first function of the Feeder Cutoff Combination is to add the small amount of water required to keep the boiler in operation. The Feeder mechanically opens its feed valve, adds the small amount of make-up water necessary, and then closes again. This allows the burner and boiler to function in the automatic and usual way.

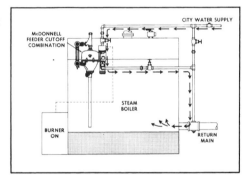

Cutoff Stops Burner
Should any condition arise where the water put into the boiler by the Feeder could not maintain the recommended water level—so that the water line dropped further—then the electrical Cutoff switch would stop the automatic burner. Notice that water is still visible in the gauge glass at the Cutoff point. As soon as the water level is restored to its lowest operating point, the electrical Cutoff switch will allow the burner to resume operation, and the system will continue to function.

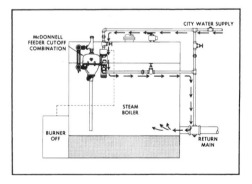

CLOSED HEATING SYSTEMS

10

Figure 13-6 **Operation of the combination feeder–low water cutoff.** *Courtesy: McDonnell Miller.*

Figure 13-7   **Probe-type low water cutoff.** *Courtesy: McDonnell Miller.*

Figure 13-8   **Electronic boiler water feed valve used with probe-type low water cutoffs.**
Courtesy: McDonnell Miller.

STEAM PRESSURE CONTROLS

Steam pressure, in a steam heating system, is controlled by the steam pressure control (see Figure 13-9). These controls are mounted on the boiler above the water level and are connected by using a pig-tail siphon (see Figure 13-10). This device ensures that there will be a cushion of water between the control diaphragm and the steam coming out of the boiler that would damage the control. As steam exerts pressure on the diaphragm or bellows in the bottom of the control, it expands, pushing against a lever that is mounted on a pivot. The other end of the lever is attached to a spring that can be adjusted to make the lever harder to move. The pivot lever is set to make a switch, either a mercury tube or a microswitch, open or close a circuit. The force or pressure required to move the lever can be adjusted by tightening or loosening the screws attached to the adjustment springs.

There are two different types of setting scales used on domestic pressure controls—additive and subtractive. When setting the additive type, you set the cut-in point first and then adjust the differential so that the two settings add up to the cut-out point. As an example, set the cut-in point at 2 psi and the differential at 3 psi. Run the system and you will see that the burner will cut off when the pressure reaches 5 psi. Now watch the pressure drop and when the pressure reaches 2 psi the burner will "cut in" again. You can use any combination of settings to reach the ideal pressure for each individual system. The cut in should never be set at 0 psi because the control may hang up and not turn the burner on. Remember that the steam safety valve opens at 15 psi so be certain to set the pressure control to cut out well below 15 psi. When setting the subtractive type of pressure control,

Figure 13-9 **Steam pressure control.** *Courtesy: Honeywell.*

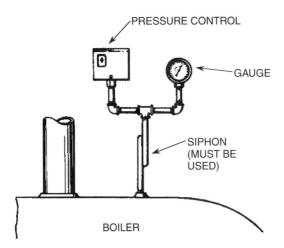

Figure 13-10 **Proper mounting location for steam pressure control.** *Courtesy: Honeywell.*

you set the "main" at the cutoff point and use the differential to determine the cut-in point. As an example, set the main at 5 psi and the differential at 3 psi. Run the system and you will see that the burner will cut off when the pressure reaches 5 psi and stay off until the pressure drops to 2 psi when it will cut in again.

The controls described above are used as oil burner limit controls because they will open the circuit when the pressure in the system rises. These are known as direct acting controls. There are other pressure controls that will close the circuit as the pressure rises and these are known as reverse acting controls. They are used to control equipment that should be turned on only when there is steam pressure in the system such as the fan motor in a unit heater. The reverse acting pressure control will ensure that the heater core is full of steam and hot before the fan motor turns on. This will guarantee the delivery of hot air to the space to be heated (see Figure 13-11).

Another type of pressure control is the manual reset pressure control (see Figure 13-12). This control is used as a back-up for the regular limit switch and is usually set to cut off at 12–13 psi. Once this control shuts off it will not go back to the closed position, even if the pressure in the system drops to 0 psi. If the manual reset pressure control shuts the system off, it is a very clear indication that there is a problem with the regular pressure control that must be corrected. This can be caused by dirt in the pig tail, a defective mercury or microswitch, or a short circuit in the wiring.

HOT WATER CONTROLS

Hot water controls, usually referred to as aquastats, are used to control the water temperature in steam and hot water boilers. The device that operates the control is a copper bulb filled with a liquid that fits snugly inside a well that is installed in the boiler below the

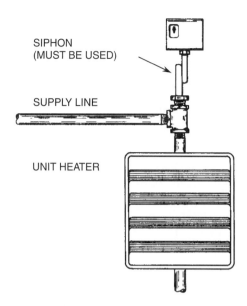

SIPHON
(MUST BE USED)

SUPPLY LINE

UNIT HEATER

Figure 13-11 **Reverse acting steam pressure control mounted at a unit heater. This control will not allow the heater fan to run until there is steam pressure at the heater. This ensures that the heater will circulate hot air.** *Courtesy: Honeywell.*

Figure 13-12 **Manual reset steam pressure control.** *Courtesy: Honeywell.*

water level. Figure 13-13 shows an aquastat with a remote bulb and capillary tubing. The bulb is connected by the capillary tubing to a bellows-type container that fits against a switch. As the water temperature in the boiler rises, the liquid in the bulb expands, exerting pressure on the bellows. The bellows moves and operates the switching mechanism, usually a microswitch.

The type of aquastat used for a limit or operating control is a direct acting switch, which means that it will open the circuit as the water temperature rises. There are also reverse acting aquastats that will close a circuit when the water temperature rises. There is also a combination control that can perform both functions (see Figure 13-14). Hydronic heating system boilers may have only one tapping for the hot water controls. This makes multiple-type aquastats necessary for the variety of possible system wiring schemes. Figure 13-15 shows a triple aquastat that can control a hydronic heating system that also delivers domestic hot water. Figure 13-16 shows the triple aquastat built into a cad cell primary control. This type of combination control is used on packaged hydronic system boilers that are assembled by the boiler manufacturers.

These various aquastats and combination controls, used either individually or in combinations, give us the ability to design a variety of control systems that will satisfy the needs of individual buildings. Chapter 14, Control Circuit Wiring, shows schematic wiring diagrams for a variety of hydronic systems. Obviously, with so many controls, there are any number of possible wiring schemes. Working in the field is really the only way to learn how to recognize and adjust hydronic systems.

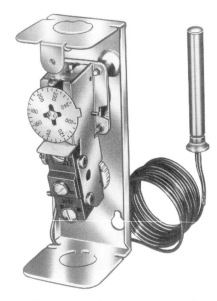

Figure 13-13 **Hot water control (aquastat) with remote bulb and capillary tubing.** *Courtesy: Honeywell.*

Figure 13-14 **Single aquastat combines direct and reverse switching functions.** *Courtesy: Honeywell.*

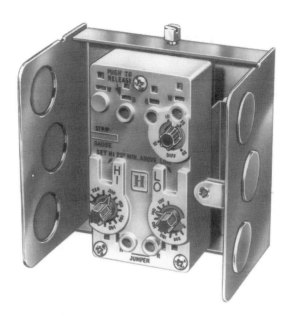

Figure 13-15 Single aquastat combines high and low limit with reverse switching function. *Courtesy: Honeywell.*

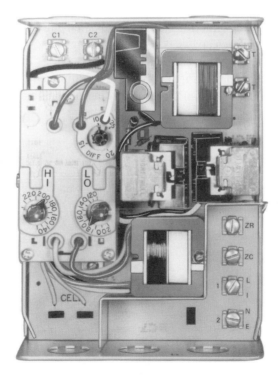

Figure 13-16 Combination triple aquastat, circulator relay, and oil burner cad cell primary control. *Courtesy: Honeywell.*

THERMOSTATS

The main operating control for any type of heating system is the thermostat. Thermostats are designed to maintain the room temperature. They do an amazingly accurate job. The operating mechanism can be a bimetal coil, a gas-filled bellows, or a semiconductor chip built into a small computer. The bimetal coil is used in most low voltage thermostats. It is extremely sensitive to temperature changes and can be attached to a small mercury tube or enclosed magnetic switch to turn the burner on when the room temperature drops a few degrees below the control setting. This type of control has a built-in heat anticipator, which is a small electric heater that operates when the thermostat circuit is closed. The purpose for the heat anticipator is to add a small amount of heat inside the thermostat cover. This will speed up the thermostat's reaction to the heat supplied by the heating system to the room and will shut the burner off. Once the thermostat circuit opens, the heat anticipator will cool off, allowing the thermostat to react to the room temperature and turn the burner on when necessary. This action of the heat anticipator helps the thermostat maintain a constant room temperature. The heat anticipator has an adjustable scale that is normally set to match the current output of the primary control. As a general rule, steam system thermostats are set for a longer "on cycle" by setting the anticipator to the "long" setting. Warm air systems require a shorter setting that is accomplished by setting the anticipator to the "short" setting.

Clocks can be built into a thermostat unit to allow us to lower the heat when it is not needed during the night or when no one is at home, and then raise it again when it is needed (see Figure 13-17). These controls can be set for one, two, or three cycles daily and different cycles for the weekend (see Figure 13-18). The timing and the temperature setting of the lowered cycle varies from building to building depending on the type of heating

Figure 13-17   **Clock thermostat has night set-back feature.** *Courtesy: Honeywell.*

Figure 13-18 **Fully programmable clock thermostat has separate weekday and weekend cycles.** *Courtesy: Honeywell.*

system and how fast the heat can be restored to the desired temperature. Considerable fuel savings can be achieved by clock thermostats that are properly set, with no loss of comfort.

The location of the thermostat is critical to the operation of any type of heating system. They should be mounted on an inside wall where they will react to changes of temperature in the living space. Be sure that there is no heat source—radiator, wall oven, warm air system register, or large lamp—near the thermostat. Any of these (and you can probably add to the list) will affect the operation of the thermostat.

FAN–LIMIT CONTROLS

Warm air heating systems utilize controls that are unique to their needs. These controls must react to changes in air temperature and humidity. The fan–limit control is installed in the furnace plenum and acts as the high limit switch, which prevents damage to the furnace, and the controlling switch for the blower section, which will not allow cold air circulation. There are two types of fan–limit controls now in use: the mechanical and the solid-state electronic controls. Figure 13-19 shows a mechanical fan–limit control. The mechanical control is fully adjustable while the solid-state control is not. The mechanical type also includes a method of running the fan continuously for air circulation when the heating system is not being used. The operation of the fan–limit control was explained in Chapter 10, Warm Air Heating Systems.

Figure 13-19 **Fan–limit control fits into a warm air furnace plenum.** *Courtesy: Honeywell.*

HUMIDITY CONTROLS

Heating air directly as we do in the warm air heating system will remove the moisture from the air. This problem has been addressed by improving the humidifiers so that they continuously add moisture to the heated air. Obviously some type of control is required to maintain the desired humidity in the building. There are several types of humidity controls. The most common is the type installed in the furnace return air duct before the air passes through the system filter. Figure 13-20 shows a combination humidity control and sail switch. The sail switch is used to activate the electronic filter. By installing the control in the return duct we can monitor the condition of the air in the building since there are additional sources of moisture present in all buildings. The installation must be done according to the manufacturer's instructions. There must be enough room for the unit inside the duct. This type of control will require maintenance and should be cleaned during the annual tune-up. There are humidity controls that mount on the wall to monitor individual rooms. They are used to control individual room humidifiers as well as central units mounted in the furnace plenum (see Figure 13-21).

PRIMARY CONTROLS

The primary control is responsible for monitoring the operation of the oil burner. This control has a flame failure safety system that will "lock out" on safety and shut the burner off if there is no flame when the burner starts. The power supply to the primary control must pass through all of the limit controls. If one of them senses an unsafe condition there

Figure 13-20 Humidity control is installed in the system return air duct to control the operation of a power humidifier. *Courtesy: Honeywell.*

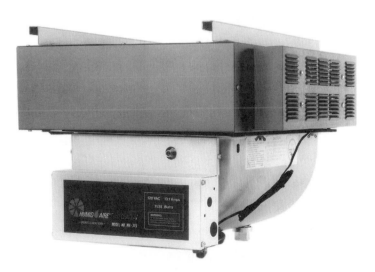

Figure 13-21 Humidifier that can be installed in the system supply duct or in an individual room. *Courtesy: Adams Manufacturing Company.*

will be no power to the primary and the burner will not be turned on. This ensures the safe operation of the system.

Primary controls require a flame detection system that will allow them to continue burner operation when there is a flame established, or lock out on safety if there is no flame when the burner attempts to start. There are two types of detection systems now in use. The older type, which is rapidly becoming obsolete, feels the heat released by the flame because it is mounted in the flue pipe between the boiler and the chimney (see Figure 13-22). When the flame is established, the hot gas released passes over a bimetal spring or helix that expands and closes a set of electrical contacts known as the *hot contacts*. Once the hot contacts are closed, the safety switch heater is taken out of the circuit and the burner can continue to operate. If there is no flame, the helix will not expand and the safety switch heater will continue to heat up until the control locks out on safety. Under normal operating conditions, the burner will shut off once the heat or hot water demand is satisfied. The helix will cool off and open the hot contacts and close the cold contacts so that the burner can be ready to start when the next call for heat or hot water is sent to the primary control. The relay, built into the primary control, will not energize if the cold contacts are open. This safety feature keeps the burner from starting if there are hot gasses in the flue pipe, indicating that there is some form of flame inside of the boiler. Figure 13-23 is a view of a stack mounted primary control with the cover removed.

Figure 13-22 **Stack mounted oil burner primary control.** *Courtesy: Honeywell.*

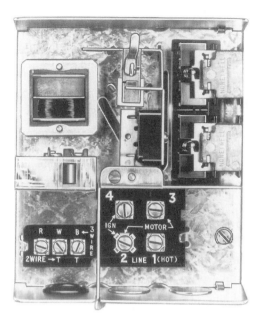

Figure 13-23 **Interior view of stack mounted primary control shows the relays, transformer, safety switch, pyrostat, and wiring terminals.** *Courtesy: Honeywell.*

Follow These Steps When Servicing Stack Mounted Primary Controls:

1. Clean the bimetal element (helix) because a build-up of soot will insulate it. This will prevent the hot flue gas from heating the helix so that the hot contacts will close, which will lead to a safety lockout.

2. Clean the cold and hot contacts. Dirt on the cold contacts will keep the relay deenergized when there is a call for heat or hot water. Dirt on the hot contacts could result in a safety lockout. Run the burner and check the action of the helix and contacts. The hot contacts must close before the cold contacts open when the burner starts. If this does not happen, the burner will shut off and start a series of short cycles resulting in no heat in the building.

3. Clean the safety switch contacts. Some stack mounted primary controls have open safety switches that may get dirty. Dirty contacts can keep the relay from energizing when there is a call for heat.

4. Check for power at the #1 and #2 terminals. #1 is the hot connection coming from the limit controls; #2 is the neutral connection. A test light connected to #1 and #2 should light up. If there is no light, check from #1 to a grounded metal surface. If the light comes on, that would indicate a broken ground wire to the control that would have to be repaired.

5. Check the low voltage circuit by installing a jumper—a short piece of wire—between the T-T terminals of the control. This will complete the circuit to the relay coil and the control should energize.

Follow These Steps to Check Out a Stack Mounted Primary Control:

1. Check for power, 120 volts, at terminals #1 and #2.
2. Check to see if the safety switch contacts are closed.
3. Check to see if the cold contacts are closed.
4. Jumper the T–T terminals. The control should energize. If it does not, it is defective and will have to be replaced.
5. Check the safety switch timing by disconnecting the #3 terminal wire to the oil burner motor. Energize the control and see how long it takes for the safety switch to lock out—30 to 45 seconds is normal for this type of control.

CAD CELL PRIMARY CONTROLS

The second type of primary control employs a flame detector that can see the flame. The most popular of these is the cadmium sulfide flame detector, commonly referred to as the cad cell (see Figure 13-24). The cadmium sulfide is formed into a grid that is enclosed in a clear plastic cover. Cadmium sulfide has a high resistance to the flow of electricity in the dark. When light strikes the cadmium sulfide grid, the resistance to electricity drops dramatically. The cad cell is installed inside the oil burner facing toward the end of the blast tube. When the burner is off, the cad cell is looking into the dark combustion chamber. On a call for heat or hot water, the primary control energizes, starting the oil burner. If there is a flame, the cad cell is looking right at it. The bright light of the oil burner flame reduces the resistance in the cad cell, allowing electricity to flow through the cell that bypasses the safety switch heater, keeping it from locking out. If there is no flame, the high resistance of the cad cell will direct the electricity through the safety switch heater that

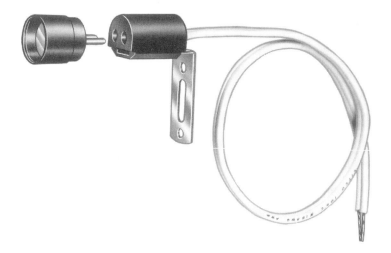

Figure 13-24 **Cad cell flame detector and wiring harness.** *Courtesy: Honeywell.*

will cause the switch to lock out. The timing of the safety switch lockout varies from 15 to 45 seconds. This information should be on a label inside the primary control cover (see Figure 13-25).

To Test the Cad Cell:

1. Remove the cell from the harness receptacle.
2. Connect an ohmmeter to the two cad cell prongs.
3. Cover the cell grid with your finger and check the ohmmeter—it should read infinity or close to it.
4. Slowly move your finger away from the grid to allow light to strike the surface and observe the ohmmeter—it should move toward the low end of the scale as more light strikes the grid.
5. Put the cell back into the harness receptacle and close the burner. Disconnect the cad cell wires from the F-F terminals and connect them to the ohmmeter.
6. With the burner off, the meter should once again indicate infinity. Run the burner and observe the ohmmeter—it should drop to somewhere between 300 and 900 ohms if the cell and the harness are in good condition and the front of the burner is not blocked.

The cad cell primary controls are usually mounted on a 1900 junction box attached to the oil burner. Some burner manufacturers have devised a junction box that can accept the cad cell relay as part of the burner assembly. The line voltage wiring to the control is connected to lead wires that extend out of the back. The black wire is connected to the hot wire coming from the limit control circuit. The white wire is connected to the neutral wire. The orange wire is connected to the oil burner motor as well as the ignition transformer on intermittent ignition systems. Interrupted ignition system primary controls

Figure 13-25 **Cad cell oil burner primary control.** *Courtesy: Honeywell.*

have a blue wire that is connected to the ignition transformer. The difference between the systems is that the intermittent system has the transformer energized whenever the motor is running, which delivers constant ignition spark at the electrode tips. The interrupted system has the ignition on for a short period of time and then, once the flame is established, the spark is shut off (see Figure 13-26).

The low voltage wiring to the control is attached to a terminal strip on the side of the unit (see Figure 13-27). The T-T terminals are where the low voltage operating control, thermostat or aquastat, is connected. The cad cell is connected to the F-F terminals. A unique feature of these primary controls is that the relay will not energize if the cad cell resistance is low, indicating that it can see light. Because the cell is inside the oil burner "looking" into the dark furnace area of the boiler or furnace, the only possible light source

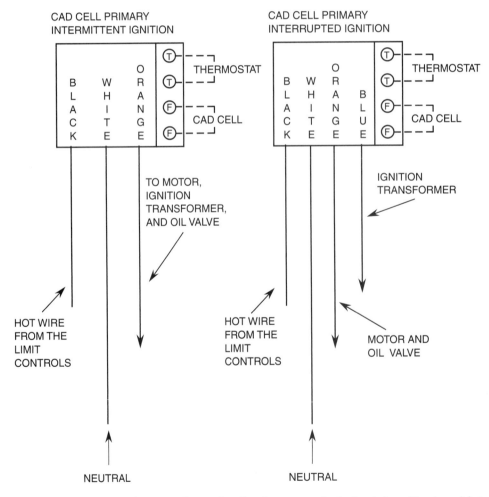

Figure 13-26 **Wiring diagrams for cad cell primary controls for intermittent and interrupted ignition systems.**

Figure 13-27 Interior view of cad cell oil burner primary control shows the relay, safety switch, transformer, and electronic components. The low voltage wiring terminals are outside of the cover. *Courtesy: Honeywell.*

would be oil burning in the combustion chamber. If there is a flame in the combustion chamber, the relay will not energize.

The technology involved in cad cell primary controls has led to the solid-state units that do not have a relay but rely on electronic switching circuits. The control shown in Figure 13-28 has all of the information concerning its operation printed on the outside cover. Additional features include an indicator light on the low voltage terminal strip that lights up to show that the control has locked out on safety. There is a manual switch that can be used to lock the control on safety as well. This is extremely handy for the service technician because the control is mounted on the oil burner and this switch can be used to shut the burner off so that it can be worked on.

Follow These Steps to Check out a Cad Cell Primary Control:

1. Check for power, 120 volts, at the black and white leads. The black lead is connected to the hot wire coming from the limit controls. The white lead is the neutral connection. If there is no power between black and white, test from black to ground. If you get a light, there is a problem in the ground connection to the control that will have to be repaired.

2. Reset the safety switch to see if the control is locked out on safety. If this does not energize the control, continue to the next step.

Figure 13-28 **Solid-state electronic cad cell primary control features an indicator that lights when the control locks out on safety due to a flame failure.** *Courtesy: Honeywell.*

3. Jumper the T-T terminals to see if there is a problem in the low voltage thermostat circuit. If the control does not energize, continue to the next step.
4. Disconnect one of the F terminal wires. This will take the cad cell out of the circuit. If the control energizes, a problem with the cad cell or the wiring to it is indicated. These primary controls have a built-in safety feature that will not allow the control to energize if the cad cell can see light. If the control does not energize, it will have to be replaced.
5. Check the safety switch timing by disconnecting the orange lead to the oil burner motor and energizing the control. The safety switch should lock out in the time, or close to the time, listed on the label.

ELECTRONIC PRIMARY CONTROLS

A new family of cad cell primary controls has evolved. They are the solid-state electronic controls developed by Carlin Combustion Technology and Honeywell. The Carlin controls have some unique features and are now standard on Carlin burners and many others as well. They offer the protection and dependability of the other cad cell primary controls and prevent overloading the combustion chamber with oil by restricting the number of times the reset button can be used by the home owner. This is because of the lockout system when there is a flame failure. The first time the control locks out it can be reset by holding the reset button down for three seconds. This is true for the second and third attempts as well. If the burner does not produce a flame after three attempts, it will lock out on safety. When this happens the reset button must be depressed and held down for thirty seconds to get the burner to run again. The home owner is not aware of this feature and will surely call for a service technician.

The Model 48245 (see Figures 13-29 and 13-30) is wired for constant ignition. It has a 45-second trial for flame and then it locks out on safety and must be manually reset. There is a red LED on the exterior low voltage wiring strip that indicates safety lockout when it is on. The thermostat and cad cell are connected to the external wiring strip. The control mounts on a standard 4″ × 4″ electrical junction box.

The Model 4020002 (see Figures 13-31 and 13-32) is wired for interrupted ignition. The trial for ignition timing is 15 seconds. The control provides an ignition timing override of 10 seconds and then the ignition shuts off. Safety switch lockout will occur after 15 seconds if there is no flame. If there is a flame failure during a running cycle, the control has a 90-second recycle time before it will attempt to start again.

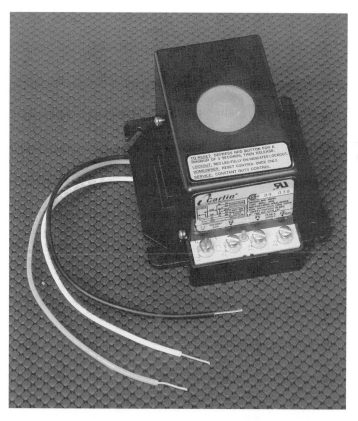

Figure 13-29 **The Carlin Model 48245 cad cell primary control.** *Courtesy: Carlin Combustion Technology, Inc.*

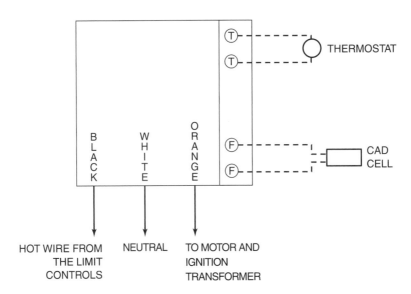

Figure 13-30   **Wiring diagram for Carlin Model 48245 cad cell primary control.**

Figure 13-31   **The Carlin Model 4020002 cad cell primary control.** *Courtesy: Carlin Combustion Technology, Inc.*

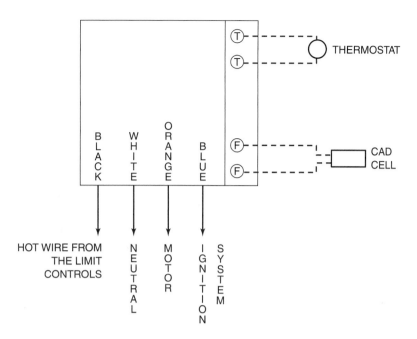

Figure 13-32   **Wiring diagram for the Carlin Model 4020002 cad cell primary control.**

The Model 6020002 (see Figures 13-33 and 13-34) is actually a program type control. This control requires a continuous supply of power to the red and white striped wire, as shown in Figure 13-34, for the internal electronic circuit. The control's sequence of operation is as follows: On a call for heat there is a 4-second self-test, indicated by a solid amber LED on the external wiring strip. This is followed by the ignition system coming on 1 second before the motor starts. The preignition timing compensates for sluggish turn on, common to some AC transformers. It is not necessary when an electronic spark generator is used. The motor then starts and runs for a 15-second prepurge period. This clears the combustion chamber of any fumes and establishes an air flow in the boiler. At the end of the prepurge, the oil valve opens and the trial for ignition period—15 seconds—begins. If there is a flame, the ignition will remain on for an additional 10 seconds and then turn off. If there is no flame, the control will lock out on safety. At the end of the normal running cycle, the oil valve will close and the motor will continue to run for 15 seconds to purge the boiler of any residual fumes. Then the motor turns off and the control recycles to get ready for the next cycle. A safety lockout is indicated by a solid red LED on the exterior terminal strip.

Honeywell's electronic primary control is the R7184 (see Figure 13-35). There are four models of this control, the A, B, P, and U, to allow some flexibility in designing a control system for a particular unit. The chart below explains the differences in these control models.

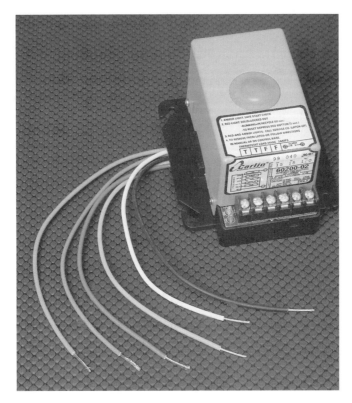

Figure 13-33 **The Carlin Model 6020002 cad cell primary control.** *Courtesy: Carlin Combustion Technology, Inc.*

R7184-A—Does not provide pre- or postpurge

R7184-B—Provides 15-second prepurge but no postpurge

R7184-P—Provides 15-second prepurge and field adjustable postpurge of 0, 2, 5, or 15 minutes

R7184-U—Provides 0 or 15-second prepurge and field adjustable postpurge

The wiring to the R7184 is connected to the bottom of the base plate by using 1/4″ quick electrical connectors. The plate is made to fit on a standard 4″ × 4″ electrical junction box (see Figure 13-36). The low voltage thermostat connections are made to external screw type terminals. Please note that a continuous supply of power must be connected to L1 for the control to operate properly. Figure 13-37 is a wiring diagram with an R7184-U programmable primary control. This control allows the installer to select pre- and postpurge timing as well as interrupted ignition—shuts off 45 seconds after the cad cell detects a flame or intermittent ignition that is on as long as the flame is on.

The LED indicator on the control provides the following information:

1. A flashing light—1/2 second on and 1/2 second off, indicates a safety lockout.
2. A flashing light—2 seconds on and 2 seconds off, indicates that the control is in a recycle mode.

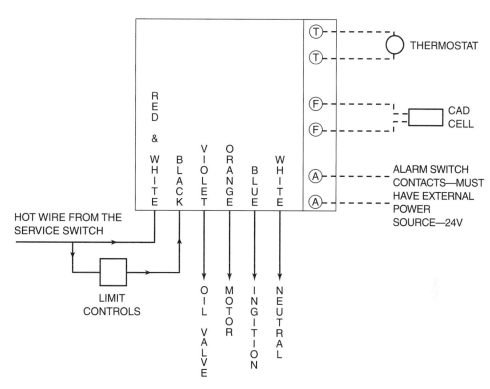

Figure 13-34 **Wiring diagram for the Carlin Model 6020002 cad cell primary control.**

Figure 13-35 **Honeywell R7184 electronic oil primary control.** *Courtesy: Honeywell.*

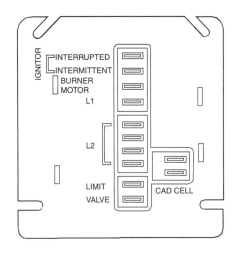

Figure 13-36 **Wiring connections in baseplate of the R7184.** *Courtesy: Honeywell.*

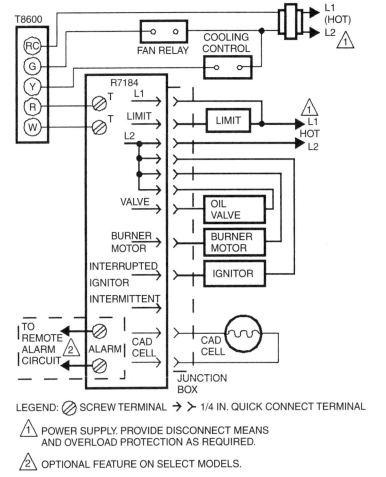

Figure 13-37 **Typical R7184-U wiring diagram.** *Courtesy: Honeywell.*

3. The light is on—indicates that the cad cell is sensing a flame.
4. The light is off—indicates that the cad cell does not sense a flame.

The LED can also be used to check the condition of the cad cell. With the burner running press the reset button. The LED will flash to indicate the resistance of the cad cell as follows:

1 flash = a resistance of less than 625 ohms.
2 flashes = a resistance of between 625 and 1,250 ohms.
3 flashes = a resistance of between 1,250 and 2,500 ohms.
4 flashes = a resistance of between 2,500 and 5,000 ohms.

To operate properly the resistance of the cad cell, when it is looking at a flame, must be below 2,500 ohms. A higher reading means that either the face of the cell is dirty or the cad cell is defective and will have to be replaced. This convenient test eliminates the use of an ohmmeter to test the cad cell.

The Honeywell R7997 (see Figure 13-38) is an integrated oil primary control and ignitor in one unit. The control is built in four models. The differences in the models are as follows;

R7997-A—No pre- or postpurge. The ignition is intermittent, stays on during the entire burner running cycle.
R7997-B—No pre- or postpurge. The ignition is interrupted, shuts off 45 seconds after the cad cell detects a flame.
R7997-C—Has a 15-second prepurge, a field adjustable postpurge and intermittent ignition.
R7997-D—Has a 15-second prepurge, field adjustable postpurge and interrupted ignition.

Figure 13-38 **The Honeywell R7997 primary control and electronic ignitor.** *Courtesy: Honeywell.*

Figure 13-39 shows how the R7997 is mounted on the oil burner chassis. The high voltage terminals must make contact with the electrode bus bars when the control is in the closed position. Care must be taken not to crimp the wires against the burner chassis when the control is moved from the open to the closed position. To facilitate the opening and closing of the control, stranded wire leads extend out of the back of the control. The color coded wires must be connected as shown in Figures 13-40 and 13-41. Please note that a constant source of power is required for the R7997 C or D but not for the R7997 A or B. The LED functions exactly as on the R7184 controls.

Any oil burner can be upgraded by installing a solenoid oil valve and one of the program type primary controls. The benefits of pre- and postpurge are smooth light off and a solution to many after drip problems because of reflected heat on the nozzle at the end

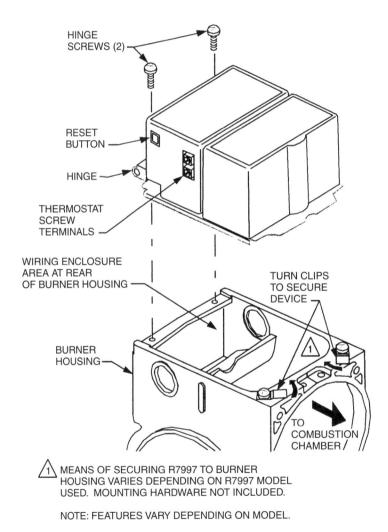

⚠1 MEANS OF SECURING R7997 TO BURNER
HOUSING VARIES DEPENDING ON R7997 MODEL
USED. MOUNTING HARDWARE NOT INCLUDED.

NOTE: FEATURES VARY DEPENDING ON MODEL.

Figure 13-39   **Mounting the R7997 on the burner chassis.** *Courtesy: Honeywell.*

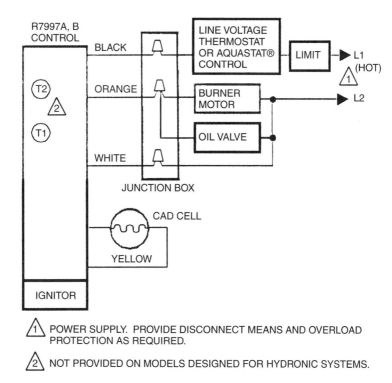

POWER SUPPLY. PROVIDE DISCONNECT MEANS AND OVERLOAD PROTECTION AS REQUIRED.

NOT PROVIDED ON MODELS DESIGNED FOR HYDRONIC SYSTEMS.

Figure 13-40 **Typical wiring diagram for R7997 A or B.** *Courtesy: Honeywell.*

of a firing cycle. Additional information on Honeywell residential oil burner controls can be found on the internet at http:/resdbbc.honeywell.com. If you want detailed information on a particular control, that can be found at http:/hbctechlit.honeywell.com.

There are several other types of primary controls that employ flame detectors that can see the flame. These are generally used on large burners and may have programming functions that control the start and shut down cycles of the oil burner. These flame detectors are:

1. The photocell that can generate a small amount of electricity when it sees the light of the flame. This small amount of current is amplified by an electronic circuit in the control to energize a relay. If this relay, known as the flame relay, does not energize, the control will lock out on safety.
2. The lead sulfide (PbS) flame detector, that generates a small amount of electricity when it sees the infrared rays in the flame. This small amount of current is amplified, by an electronic circuit built into the control to energize a flame relay. If the flame relay does not pull in, the control locks out on safety. PbS flame detectors are used primarily for multifuel burners or for burners that use gas pilot type ignition systems. The PbS cell can "see" the infrared rays in a gas flame to prove the pilot before the main oil valve is allowed to open.

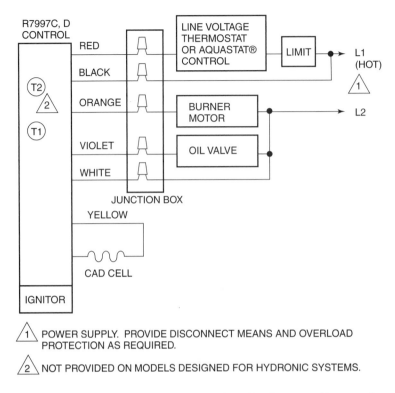

Figure 13-41 **Typical wiring diagram for R7997 C or D.** *Courtesy: Honeywell.*

3. The ultraviolet (UV) flame detector can see the ultraviolet rays in the oil burner flame. The same type of electronic circuit is used to energize a flame relay. If there is no flame, and the flame relay does not pull in, the control will lock out on safety. UV flame detectors can be used for multifuel and gas pilot ignition burners because they can "see" the ultra violet rays in gas and oil flames.

The primary control acts as a relay to start the burner when one of the operating controls calls for heat. The relay coil circuit is powered by a step down transformer built into the control. The power supply to this transformer must pass through all of the limit controls that ensure safe operation of the system. The use of low voltage relay coils to operate the oil burner allows us to use small, sensitive operating controls. Many oil burner technicians refer to the primary control as the "relay."

SWITCHING RELAYS

Switching relays are used to allow a low voltage control to operate a line voltage appliance (see Figure 13-42). The most common use for switching relays is in hydronic heat-

Figure 13-42 **Switching relay commonly used to operate circulators in hydronic heating systems.** *Courtesy: Honeywell.*

Figure 13-43 **Combination triple aquastat and circulator relay.** *Courtesy: Honeywell.*

ing zoned systems. Individual relays are connected to operate the zone circulators. Figure 13-43 shows a combination triple aquastat and switching relay. Newer versions of these controls have base panels that can accommodate many plug-in relays. Regardless of type, these units all have a step down transformer to supply the power for the low voltage operating control circuit.

REPLACING CONTROLS

Replacing any one of the controls once it has proved to be defective must be done properly. The best replacement is an exact replacement. All the connections are the same, electrical as well as piping, so the job can be done perfectly. If you do not have an exact replacement, you must be certain to make the installation according to the local building code. The wiring must be properly connected using the appropriate conduit or cable and connectors. If the control has a piping connection, it must be done according to the manufacturer's instructions. Control manufacturers invest a lot of money in creating units that are attractive as well as efficient. The completed installation should look as though it was part of the original installation.

To Briefly Summarize

1. An oil burner control is an automatic switch that includes a mechanism that can react to changes in liquid level, pressure, liquid temperature, air temperature, humidity, or air movement.
2. Limit controls are used to keep the system operating within safe temperature, pressure, and water level limits.
3. Operating controls turn the burner on when there is a call for heat or hot water.
4. Primary controls monitor the oil burner flame and lock out on safety if there is no flame when the unit starts. The primary control power is supplied by the limit control circuit. The relay coil energizes when there is a complete circuit through one of the operating controls.
5. Replacing controls must be done in accordance with the local building codes and the control manufacturer's instructions.

Please Answer These Questions

1. What is a limit control?
2. What is an operating control?
3. What is a primary control?
4. How does a stack mounted primary control monitor the oil burner flame?
5. How does a cad cell primary control monitor the oil burner flame?
6. Explain the procedure for checking a cad cell flame detector.
7. Explain the operation of a combination low water cutoff water feeder.
8. Explain how an electric water feed valve is controlled by a low water cutoff.
9. What is the function of the manual reset pressure control?
10. Explain the difference between direct acting and reverse acting controls. How are they used?
11. Why are dual and triple aquastats an important part of hydronic heating systems?
12. Explain the function of a humidity control in a warm air heating system.
13. Explain how it is possible to run the blower section of a warm air system when there is no call for heat.
14. Why can a PbS or UV flame detector be used to monitor multifuel burners?
15. Explain the function of a switching relay in a hydronic heating system.

CHAPTER 14

CONTROL CIRCUIT WIRING

Control circuits for heating systems are designed to make the systems operate safely and efficiently. This is accomplished by assembling controls that can monitor the potentially dangerous parts of the system as well as the conditions that require the operation of the system. Designing a control system is an exercise in logic. Certain general rules apply regardless of the type of system.

1. All electrical wiring must comply with either the local electrical code, if there is one, or the National Electrical Code. Many areas require that electrical work be done by licensed electrical contractors. It is to your advantage as an oil burner technician to be able to wire an oil burner control circuit.
2. Every oil burner circuit must start from a dedicated fuse or circuit breaker, usually 15 amperes for residential oil burners. Oil burners are not plug-in appliances and must be permanently wired. There should be no other appliances connected in this circuit.
3. There must be a remote emergency switch that is clearly marked and separated from any other switches. This switch should be at the entrance to the boiler room or basement so that if there is a problem with the burner it can be shut off from a safe distance.
4. It is a very good idea to install a switch in the boiler room to be used by technicians working on the oil burner. This can be referred to as a service switch and the boiler controls should be wired after this switch.

The controls are divided into three basic categories: the limit controls, the operating controls, and the primary controls. In most residential systems, the limit controls are connected in the 120-volt line while the operating controls, which are smaller and more sensitive, are connected to the 24-volt circuit supplied by the primary control. The primary control receives its power from the limit controls, starts the burner on a signal from the operating controls, and monitors the burner to ensure safe operation. The limit controls are always wired in series with the fuse, remote control, and burner service switches. Because an opening anyplace in a series circuit will shut down the entire circuit, an unsafe condition or open switch will shut the burner off. If there are two or more operating controls they must be wired in parallel so that any one of them can turn the burner on.

RESIDENTIAL STEAM SYSTEM WIRING CIRCUIT

A typical, residential steam system would have the following limit controls:

1. A steam pressure control to maintain the desired system pressure.
2. A low water cutoff to shut the burner off if there is not enough water in the boiler. This will prevent serious damage to the boiler.

   The same system would have these operating controls:

1. A thermostat located in the living space of the building. The location of the thermostat is extremely important. It should be in a large room, on an inside wall, where it can truly monitor the temperature in the living space.
2. A hot water control mounted in the boiler to provide domestic hot water or to keep the boiler water from cooling off too much between heating cycles.

A Simple Steam System Wiring Circuit Operates as Follows (Figure 14-1):

1. When all of the switches are on and the low water cutoff senses sufficient water in the boiler, the aquastat turns the burner on to maintain minimum boiler water temperature. If the boiler produces domestic hot water, the aquastat would be set at 180°F. If the boiler does not produce domestic hot water, the setting could be lower.
2. When the thermostat calls for heat, the burner starts. The thermostat overrides the aquastat because they are wired in parallel. The boiler starts to produce steam and the steam enters the building radiation.
3. The steam pressure control will stop the burner if the pressure exceeds its setting even though the thermostat is not satisfied. This is not a problem because heat will

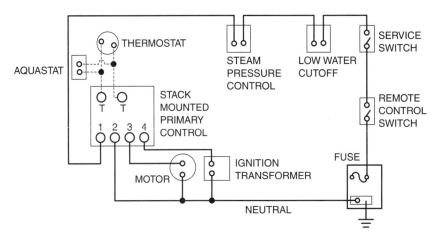

Figure 14-1 **Residential steam system.**

continue to be delivered to the building because all of the radiators are hot when the pressure builds up.

4. Once the thermostat is, satisfied the burner shuts off and all of the pressure in the system drops to zero.

Local codes may require additional controls such as a manual reset pressure control. This control is used to back up the system pressure control. It is usually set at 10 psi and once it shuts off it has to be manually reset to get the burner back into operation. This control is wired in the limit control circuit. It is important to remember that if it has to be reset, you have to check the system pressure control to find out why it did not shut the system off at its set maximum pressure. Figure 14-2 shows how these controls fit into the circuit.

An auxiliary water feed valve can be installed to add make-up water to the system. The wiring circuit for these valves is shown in Figure 14-3.

Some steam systems employ unit heaters instead of, or in addition to, the radiators. A unit heater has a steam coil and a fan. When the steam coil is hot, the fan turns on and blows air through the coil, heating the area downstream from the heater. These control circuits usually include a switch, a line voltage thermostat, and a reverse acting pressure control. The reverse acting pressure control will close the circuit when the steam pressure in the piping to the unit heater rises. This ensures that the steam coil in the unit heater will be hot before the fan turns on. Figure 14-4 is a typical unit heater control

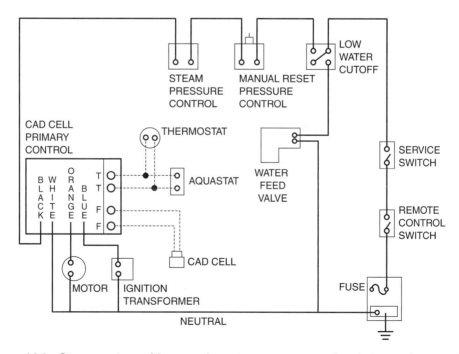

Figure 14-2 Steam system with manual reset pressure control and electronic water feed valve.

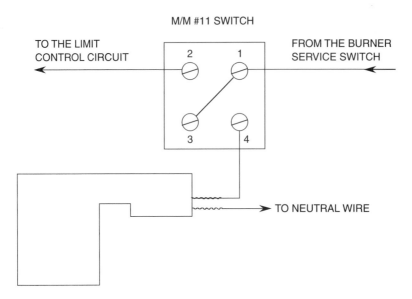

Figure 14-3   Wiring a low water cutoff to operate an electric boiler water feed valve.

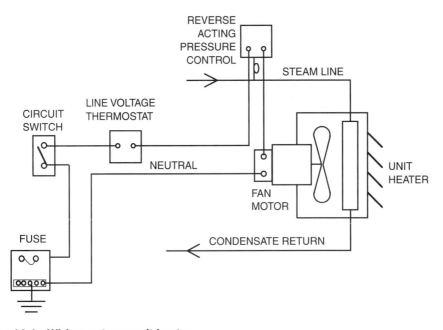

Figure 14-4   Wiring a steam unit heater.

circuit. A variation of this is to install a strap-on type of reverse acting hot water control on the return line from the unit heater. Once this pipe heats up, indicating that the coil in the heater is hot, the switch closes and the fan comes on. The wiring is done exactly the same as when the pressure control is used.

RESIDENTIAL HOT WATER SYSTEM WIRING CIRCUITS

There are many variations to the control circuits for hot water heating systems. Regardless of how the system operates there must be at least one line voltage hot water control wired as a limit switch. Some of the older gravity feed systems used a low voltage control for this purpose. This can be dangerous because if the primary control relay armature were to stick in the closed position, the low voltage limit could not shut the burner off. Figure 14-5 shows a gravity hot water system that includes a line voltage limit control.

Forced circulation systems can be divided into two categories, integrated and nonintegrated. Figure 14-6 shows a nonintegrated system. Nonintegrated systems have separate control of the circulator and oil burner. This nonintegrated system has a dual aquastat that acts as the high limit control for the burner and the reverse acting control for the circulator. A switching relay in the circuit allows us to use a low voltage thermostat.

A Nonintegrated Forced Circulation System Operates as Follows:

1. There is a probe-type low water cutoff included in this circuit. Some local codes require a low water cutoff on hot water heating systems. This is especially important

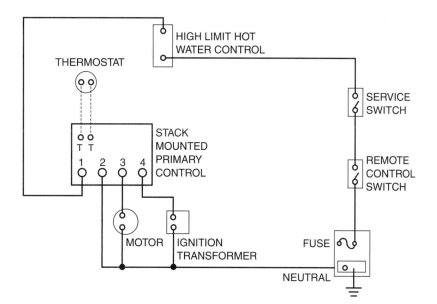

Figure 14-5   **Gravity hot water system.**

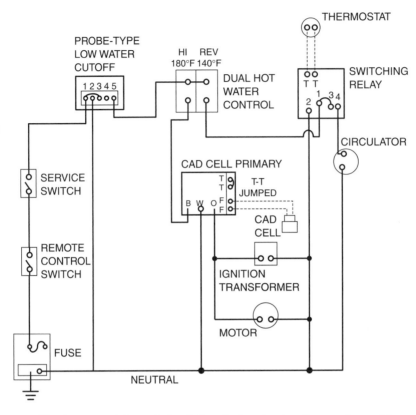

Figure 14-6 **Nonintegrated forced circulation hot water heating system with probe-type low water cutoff required when radiation is below boiler.**

when portions of the system piping are below the boiler, which can happen in a radiant heating system. An underground leak could drain the water out of the system. The low water cutoff will protect the boiler if this happens. Please notice that this control is the first control in the circuit. It is important to wire the circuit this way because the probe-type low water cutoff requires a constant power supply to operate properly (see Figure 14-7).

2. The high limit control maintains the boiler water temperature at 180°F.

3. When the thermostat calls for heat, the switching relay is energized, which starts the circulator. Hot water from the boiler is now pumped into the system radiation and cold water from the system returns to the boiler.

4. As cold water enters the boiler, the high limit switch closes starting the oil burner when the water temperature drops to approximately 170°F.

5. In systems that contain a large amount of water, it would be possible for the burner to be unable to heat the water as it circulates through the boiler. The reverse acting switch will shut off the power to the switching relay when the water temperature drops below 140° F, which will stop the circulator and give the burner a chance to heat the

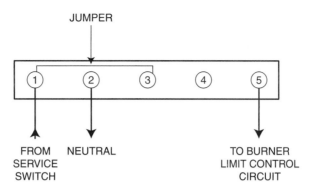

Figure 14-7 **Wiring connection for McDonnell Miller PS851 probe-type low water cutoff for hot water heating system.**

water in the boiler. This is absolutely necessary when the boiler is providing domestic hot water.

6. As warmer water returns to the boiler, the circulator running periods get longer until the thermostat is satisfied and the switching relay is deenergized, stopping the circulator.

7. The burner will continue to operate to get the water temperature back to 180°F. Please note that the T-T terminals on the primary control must be jumped for the burner to start. A common variation of this circuit includes a low limit aquastat wired in the low voltage primary control circuit.

An integrated circuit has the burner start when the circulator relay is energized by the thermostat. Figure 14-8 shows one version of an integrated circuit. There are three separate aquastats in this circuit. The high limit control is wired in the line circuit to the primary control and is set to open at 200°F. The reverse acting aquastat controls the power to the switching relay and is set to open at 140°F. The low limit aquastat is wired in the primary control low voltage circuit and is set to open at 180°F. The thermostat is wired to the T-T terminals of the switching relay and will energize this control on a call for heat. An auxiliary set of contacts in this switching relay will close when the control is energized. These contacts are wired in parallel with the low limit aquastat in the primary control low voltage circuit.

An Integrated Forced Circulation System Operates as Follows::

1. The low limit aquastat maintains the boiler water temperature at 180°F.

2. When the thermostat calls for heat, the switching relay will energize and start the circulator. At the same time, the auxiliary switch in the switching relay closes and starts the burner, overriding the low limit control.

3. If the water temperature in the boiler drops below 140°F, the reverse acting aquastat will shut off the switching relay. The burner will continue to operate because the low limit aquastat switch will be closed. When the water temperature rises above 140°F,

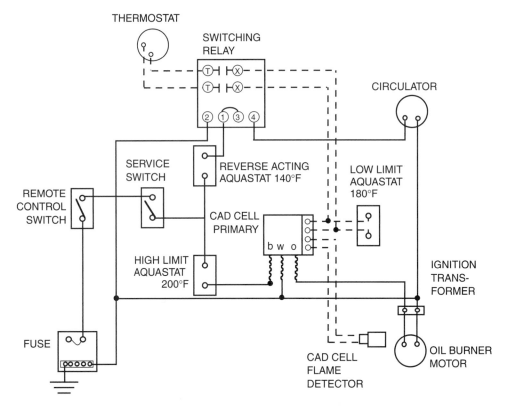

Figure 14-8   **Integrated forced circulation hot water heating system.**

the reverse acting aquastat switch will close, sending power to the switching relay and starting the circulator again.

4. If the water temperature rises above 200°F, the high limit control will open and shut the oil burner off while the circulator continues to operate. The process continues until the thermostat is satisfied.

5. When the thermostat opens, the low limit aquastat gains control of the system again and returns the water temperature to 180°F.

6. If a low water cutoff is required it must be the first control in the circuit after the service switch. This is because the control needs a constant power supply to operate properly.

The three aquastat and switching relay system described above has now been replaced by a single control. This is made necessary for hydronic boilers that have only a single tapping for the aquastat. Figure 14-9 shows the internal wiring diagram for the Honeywell L8124 triple aquastat relay.

Splitting the system into separate zones can be accomplished either by using zone valves or separate circulators. Figure 14-10 shows how zone valves may be wired. This

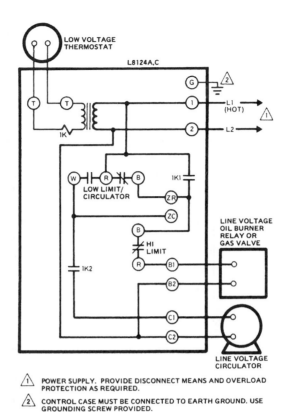

Figure 14-9 Single-zone hydronic heating system using a combination triple aquastat and circulator relay. *Courtesy: Honeywell.*

circuit includes the Honeywell R8182D that has a triple aquastat, switching relay, and cad cell primary control all in a single unit. The triple aquastat should be set as follows:

High Limit—200°F
Low Limit—180°F
Differential (boiler provides domestic hot water)—10°F
Differential (boiler does not provide domestic hot water)—25°F.

The differential setting affects only the reverse acting portion of the triple aquastat.

The power supply for the zone valves is a single step down transformer. The transformer must be capable of supplying power to all three valves at the same time. Check the instructions packed with the zone valves carefully to make sure you are using a transformer that can handle the job. Each zone valve has an end switch that closes a circuit when the valve opens. The end switches in this circuit are wired in parallel to the primary control low voltage circuit.

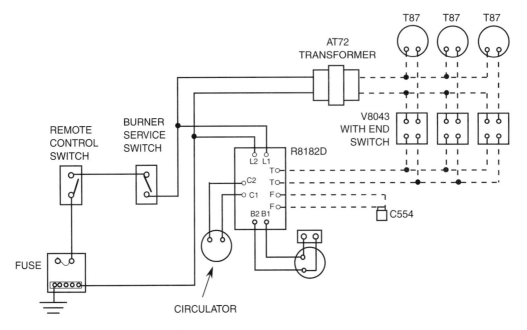

Figure 14-10   **Three-zone hydronic heating system using zone valves.**

This Is How This Circuit Operates:

1. The boiler water temperature is maintained at 180°F by the low limit switch in the R8182D.
2. When one of the zone thermostats calls for heat, its zone valve opens and the end switch in the valve closes, completing the low voltage circuit in the R8182D. This turns on both the oil burner and the circulator.
3. Hot water is sent to the zone calling for heat. The triple aquastat controls the operation of the circulator and burner.
4. When the zone thermostat is satisfied, the zone valve closes, stopping the circulator. The burner runs until the low limit setting is reached.

Zoning can also be accomplished by using individual zone circulators. Figure 14-11 shows how this type of system may be wired. This circuit uses a dual aquastat for high limit and reverse acting functions. A low limit aquastat could be added and wired to the primary control low voltage circuit to replace the jumper between the T-T terminals. The three individual switching relays shown in the diagram may be replaced by a single unit containing three plug-in relays.

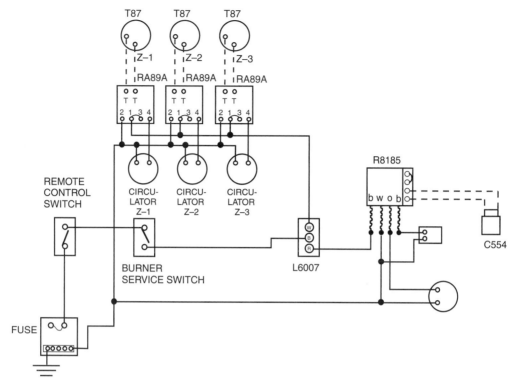

Figure 14-11   **Three-zone hydronic heating system using separate circulators.**

This Is How This Circuit Operates:

1. The water temperature in the boiler is maintained by the high limit portion of the aquastat.
2. When one of the zone thermostats calls for heat, its switching relay is energized and the zone circulator sends hot water to the zone.
3. The dual aquastat controls the operation of the burner and circulator.
4. When the zone thermostat is satisfied, the switching relay opens and stops the circulator.
5. The burner runs until the high limit temperature is reached.

It is not practical to attempt to include all of the possible variations in hydronic system wiring. The circuits illustrated here are basic and function very well. Once you understand the function of an individual control you can adapt it to achieve a particular result. Changing controls will affect the operation of the system. Try to keep any control system as simple as possible and do not arbitrarily make changes to systems that are operating properly.

WARM AIR SYSTEM WIRING CIRCUITS

There is a great similarity in the control circuit wiring of warm air and hot water systems. The controls are different but many of the wiring hook-ups are identical. Figure 14-12 is a basic circuit for a simple forced warm air system. The "brain" in this circuit is the fan-limit switch. This unit is mounted in the furnace plenum and controls the operation of the burner and the blower section. The setting should be approximately 130°F as a high limit, and a differential setting of 30°F. The high limit portion of the control is a direct acting switch that will open on temperature rise. The differential relates only to the reverse acting portion that controls the blower section. The idea is to ensure that only warm air is circulating to heat the building.

This Is How This Circuit Operates:

1. When the thermostat calls for heat, it completes the primary control low voltage circuit and the oil burner starts.
2. When the temperature in the furnace plenum reaches the reverse control setting, the blower starts and warm air is sent into the building.

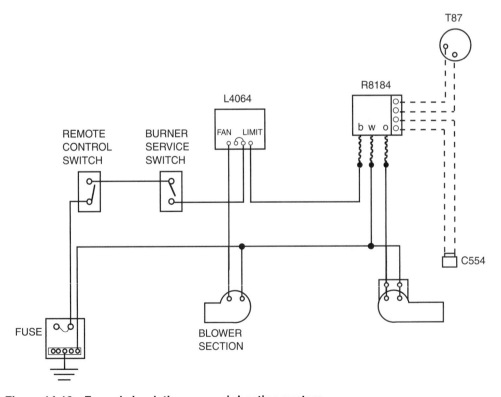

Figure 14-12   **Forced circulation warm air heating system.**

3. If the air leaving the furnace plenum cools off, the blower will stop to give the burner a chance to "catch up." This may result in short cycling of the blower, which can be avoided by slowing it down.

4. Once the thermostat is satisfied, the burner will stop, but the blower continues until the air in the plenum cools off. This leads to some overheating, which can be controlled by adjusting the thermostat heat anticipator to the fastest setting.

Figure 14-13 is an upgraded version that includes a power humidifier, a humidity control to monitor it, and an electronic filtering system. Please note that both the humidifier and the filter are on only when the blower is operating. Zoning can be accomplished by installing motorized dampers in the supply ducts to various portions of the building. Each of the motorized dampers has an end switch that would be wired to the primary control low voltage circuit. The zone thermostat would open the damper, which would start the oil burner; this would heat the air in the plenum to start the blower, humidifier, and electronic filter.

It is sometime necessary to supply constant power to one of the circuit components. This is true for the nozzle line heater that is on all the time. Figure 14-14 shows how to

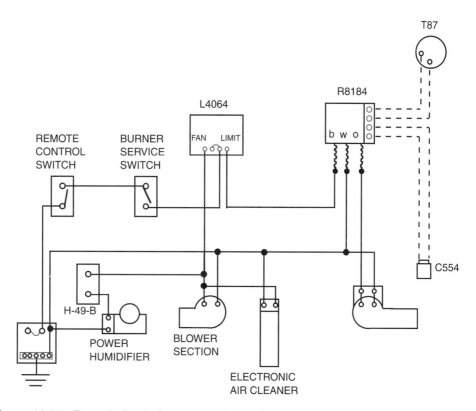

Figure 14-13 Forced circulation warm air heating system with a power humidifier and electronic air cleaner.

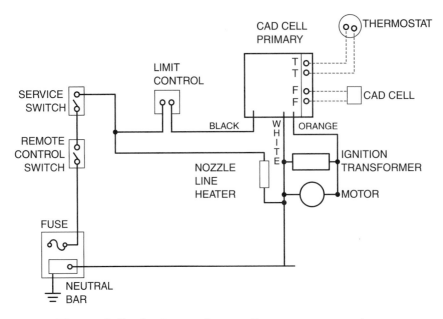

Figure 14-14 The nozzle line heater requires continuous power supply.

connect the heater in a simple oil burner circuit. Please note that the heater is powered directly from the service switch. The service switch is now more important than ever because it allows you to shut the power to the heater off when you are removing the burner drawer assembly. Constant power is also required for the program type primary controls, either the Honeywell or the Carlin. Figure 14-15 shows the Honeywell R8184P installed in a hot water system that also has a power venting system. Notice that the black primary control lead is connected directly to the power before the limit controls. To complete this diagram, you would have to show the fuse, remote control, and service switches connected to the L 1 hot wire shown on the diagram. The same is true for the heating–cooling system shown in Figure 14-16.

Riello oil burners require a switching relay (see Figure 14-17) if a low voltage thermostat is used.

To Briefly Summarize

1. All wiring must comply with the local building or electrical code.
2. Control circuit wiring is an exercise in logic.
3. Controls are assembled to ensure safe, efficient operation of the heating system.
4. The fuse, burner switches, limit controls, and primary control are all wired in series to the hot wire.
5. The system operating controls are small, sensitive switches that are wired in the primary control low voltage circuit.
6. The best control circuit regardless of type of system is the simplest hook-up that can achieve the desired results.

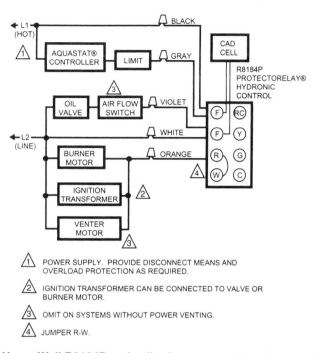

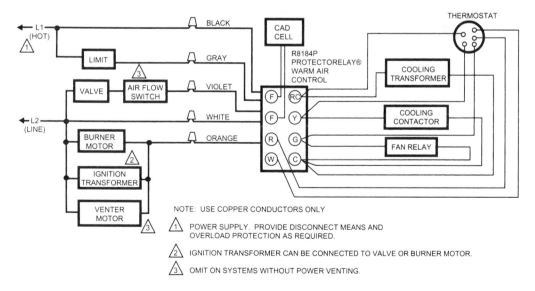

Figure 14-15   **The HoneyWell R8184P cad cell primary control requires continuous power supply to black lead wire. This is a hot water system with a power venting system.** *Courtesy: Honeywell.*

Figure 14-16   **A heating–cooling system with an R8184P program type primary control requires constant power supply to the black lead.** *Courtesy: Honeywell.*

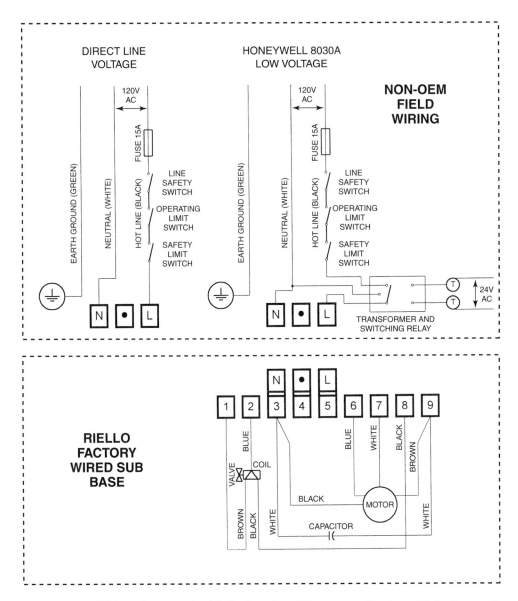

Figure 14-17 **Wiring diagrams for Riello F series oil burners.** *Courtesy: Riello Corporation of America.*

Please Answer These Questions

1. Where does every oil burner circuit start?
2. Why is it necessary to install a remote control emergency switch? Where should it be installed?
3. How can a valve be wired into a steam system circuit to add make-up water to the system?
4. How are limit controls wired to ensure that any one of them can shut the burner off?
5. Why does the hot water high limit control have to be wired in the line voltage portion of the control circuit?
6. What is the difference between an integrated and nonintegrated hot water system?
7. What is a reverse acting control? Give an example of how it is used.
8. What does the term *zoning* mean and how can it be accomplished in a hot water system?
9. Explain how you would set the differential setting of a triple aquastat on a hydronic system that produces domestic hot water.
10. Explain the sequence of operation of a warm air system.
11. What is the function of the humidity control and where is it installed?
12. How can we be sure that the electronic air cleaner is on only when the blower motor is operating?
13. Explain the operation of a two-zone warm air heating system.

CHAPTER 15

SERVICE PROCEDURES—BURNER NOT OPERATING

The preceding chapters have dealt with the operation of the various systems involved in oil-fired residential heating. Hopefully, you have been exposed to shop or field experience that will help you understand the equipment you will be working on. This chapter and the following chapters deal with the actual work of servicing and maintaining oil-fired heating systems.

The service department of any fuel oil dealer or oil burner company has the responsibility of repairing their customer's equipment when it is shut down or operating improperly. When a complaint is called in to the service department, a technician is dispatched to the job to make the necessary repairs. The response time is a major factor in maintaining the customer and recruiting new accounts. Once the technician is on the job, it is equally important to identify the cause of the problem and repair it as quickly as possible. In order to accomplish this task, the technician must follow a systematic routine that will lead to the solution of the problem. The charts that follow are for your use. They are an organized system designed to assist you in identifying and solving service problems.

Where do you start? If the homeowner is there ask questions, start a conversation, ask:

1. Is the burner running at all?
2. Was there any noise or vibration before it stopped?
3. Was there any odor or smoke?
4. When did you get your last oil delivery?

This conversation should take place while you are walking with the customer to the boiler or furnace room. On your way, you must have passed the remote control switch. Was it on? Did you pass the oil tank? Did you check the gauge to see if there is oil in the tank? Are the lights in the basement on? OK, the lights are on indicating there is power in the building. The remote control switch is on, there is oil in the tank, and the boiler is ice cold. You are face to face with a Burner Not Operating (BNO) service call.

The BNO service calls are due to one of the following possibilities:

1. The burner attempted to start but there was no flame and the primary control locked out on safety.
2. There is no power to the primary control.
3. There is an open low voltage circuit to the primary control.
4. There is a problem with the flame detector.

How do you find out into which category this particular problem falls? Open the boiler or furnace door and look inside. Is the combustion chamber in good condition? Is it wet inside? Can you smell oil? Shut the power off at the most convenient switch. Reset the primary control safety switch. Stand off to the side of the boiler in a spot where you can still look into the combustion chamber. Turn the power back on. If the burner runs, whether there is a flame or not, you can assume that it had tried to start and there was no flame and then try to determine why. If the burner does not start, you can eliminate all the things you would have to do if it did start and concentrate on the power supply first, then the low voltage circuit, and finally the flame detector. The cause of the shut down will show up during this process. You have to recognize the problem and make the repair.

All of the charts dealing with BNO service calls start at the reset switch of the primary control. This is clearly the most logical place to start. Before you press the reset switch, find a convenient place to shut off the power to the burner. This is necessary because you will want to stop the burner quickly if there is no flame. Open the boiler door so that you can see what happens when you reset the safety switch and the burner starts. Be sure to stand off to one side. Do not stand directly in front of the open boiler door. Shine your flashlight into the combustion chamber and if there is an oil spray, you will see it. You will be able to hear the ignition spark, which makes a sizzling sound. You may want to use a flame mirror to check the ignition. If you follow the charts and do not get off the track, you will surely learn how to service oil burners.

USING CHART I. BNO—NOT LOCKED OUT ON SAFETY

The primary control locks out on safety if the oil burner attempts to start and there is no flame. This chart (Table 15-1) deals with burner shut downs that have nothing to do with the flame. All of the possible causes of nonsafety shut down are listed on the chart. They form individual tracks that can be quickly identified and followed to a satisfactory conclusion. Please be aware that when you try something, such as resetting the safety switch, there are always several possible reactions to your action. You have to recognize the significance of each reaction to what you have done. This will ultimately lead you to the solution of the problem.

You have reset the safety switch, all of the switches are on and the burner is not operating. The first step on the chart is to check for power at the primary control. This is done by connecting a voltmeter or test light to the 1 and 2 terminals of a stack relay or to

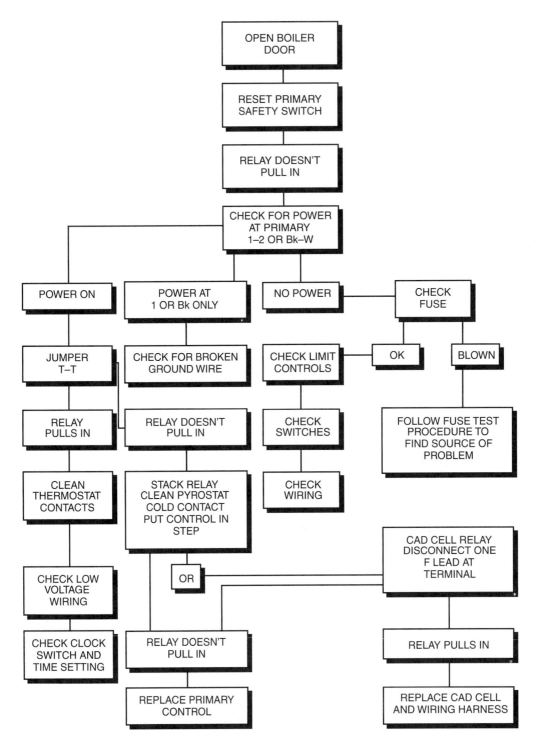

Table 15-1   **Chart I. BNO—Not Locked Out on Safety.**

the black and white leads of a cad cell relay. If your voltmeter indicates 120 volts, you can be absolutely certain that the fuse is good and all of the limit controls are closed and ready to go. If there is no voltage indicated, immediately disconnect the voltmeter lead from the 2, or white, lead and touch it to a grounded surface. If you now get a voltage indication, the problem is a broken or disconnected ground wire in the oil burner circuit. If there still is no voltage indicated, you must check the fuse, all of the switches, and limit controls to find the cause of the shut down. Check for power at each switch with a voltmeter or test light. If one of the limit controls has power coming in and nothing going out, you can assume that there is something wrong at that particular control. You must make the required repair to get the burner operating. NEVER JUMP OUT THE SWITCH ON A LIMIT CONTROL! The limit switches each monitor a potentially dangerous or damaging operation of the boiler or furnace. Jumping the switch will eliminate the safety feature the particular control is supposed to provide, and you will be responsible for any damage that results.

If there is power at the primary control and the burner still will not operate, we have narrowed the possible causes down to the low voltage circuit, the flame detector, or a defective primary control. To check the low voltage circuit, place a jumper wire across the T-T terminals. If the burner starts, the problem is in the low voltage circuit, either the control switches or the wiring. If the burner does not start, we have to check the flame detector as indicated on the chart. If the burner still does not start, the primary control has to be replaced.

The procedure on Chart I takes only five minutes to complete. The actual repair may take a considerable amount of time but you will not have wasted additional time in checking items that are totally unrelated to this particular category of BNO service call.

USING CHART II. BNO—LOCKED OUT ON SAFETY—MOTOR

Chart II (Table 15-2) is one of two charts that deal with problems with the oil burner motor. If the burner primary responds to a call for heat by energizing and the motor does not run, the primary control will lock out on safety. When you reset the safety switch and turn the power on, you will see or hear the relay armature snap into the closed position. If the motor does not run you must reset the motor overload switch. If this starts the motor, you have to follow the procedure outlined in Chart III (Table 15-3). If the motor does not start, you have to check for power at the 3, or orange, lead wire of the primary control. This must be done quickly because the control will lock out on safety again if the burner is not running. If there is no power at the primary terminal, you will have to replace the primary control. If there is power at the primary terminal, you must check for power at the motor. If there is no power, check the wiring between the primary and the motor. If there is power, you will have to replace the motor. Please remember that the primary control safety switches are heat actuated and if you keep resetting them, the lockout timing

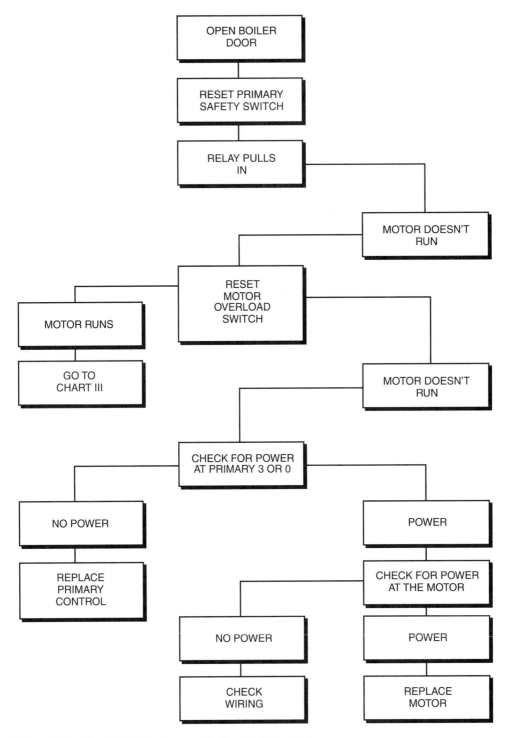

Table 15-2   **Chart II. BNO—Locked Out on Safety—Motor.**

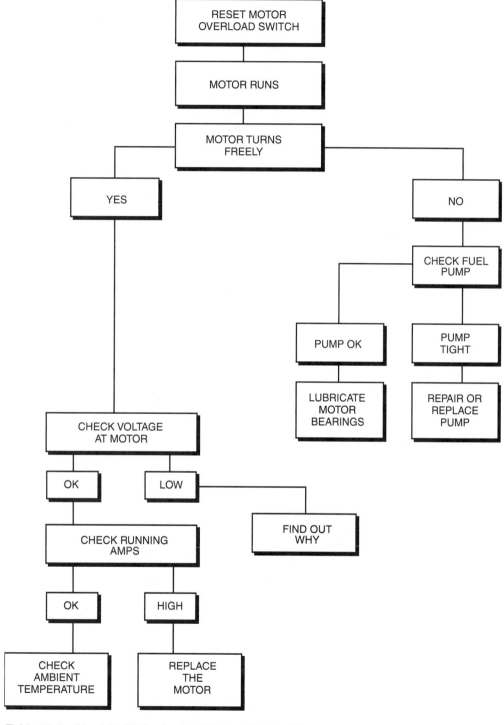

Table 15-3 **Chart III. BNO—Locked Out on Safety—Motor.**

gets shorter and shorter. If you do not work quickly, you may be testing for power at the motor with the primary locked out!

Always try to make exact motor replacements. Check the motor nameplate for the size in hp, the speed in rpm, and the running ampere rating. Check the rotation of the motor before you mount it on the oil burner. Most of the oil burner motors now in use can be reversed by switching two wires at the starting switch. Instructions for changing the rotation are either on the motor nameplate or on the plate that covers the starting switch assembly. The motor has to fit on the burner with the bearing lubrication openings facing up so that they can be properly lubricated. The wiring connections must be done according to the local electrical code.

USING CHART III. BNO—LOCKED OUT ON SAFETY—MOTOR

If the burner attempts to start and the motor overload trips, there will be no flame and the primary safety switch will lock out. The motor overload switch protects the motor and keeps it from burning out. Once the overload switch trips out, you must find out why. Just resetting the switch to get the burner running is not the solution to this service problem.

The first thing we have to determine is, does the motor turn freely? Turn off the power to the burner and open the transformer or burner cover plate so you can turn the fan and motor shaft by hand. How does it feel? If it turns freely, you can assume that the motor bearings are lubricated and that there is nothing wrong with the fuel pump that is attached to the motor shaft by the oil pump coupling. You now have to check the power supply to the motor. If the voltage is low, the motor will draw more amperes to get the watts required for its operation. You have to find out why the voltage is low. Is it a problem in the building? Is it a neighborhood problem? Is there a defective switch or splice in the oil burner circuit? All of these are potential causes of reduced voltage at the motor. Check the running amperes of the motor. They should match the ampere rating on the motor nameplate. High running amperage can be caused by an undersized motor or possibly by a defective fuel pump. You can check the fuel pump by removing the motor from the burner and turning the pump shaft by hand. A tight or hard to turn pump shaft could be caused by rusted pump gears due to water in the oil. The pump may have to be replaced to correct the condition. Another cause of high running amperage is a lack of lubrication in the motor bearings. Use a few drops of lubricating oil in each bearing and run the motor for a while. If the running amperage returns to the nameplate rating, you have solved the problem.

Another possible cause of motor overload trip out is excessively high ambient temperature. Run the burner for a while to see how hot the motor gets. Check the nameplate for a temperature rise indication. Some motors are capable of running at high temperatures; some are not. If the motor is getting hot, it may be due to a lack of insulation in the front of the combustion chamber. See if the front plate of the boiler is getting hot. Be sure there is an adequate supply of fresh air entering the boiler room. It may be necessary to place some insulating material between the motor and the front of the boiler.

USING CHART IV. BNO—LOCKED OUT ON SAFETY— NO IGNITION

The safest way to approach any BNO is to assume that the burner attempted to start and there was no ignition to start the flame. This would result in oil spraying into the combustion chamber for the length of time it takes for the safety switch to lock out. The oil will remain soaked into the combustion chamber or on the surface of the boiler where it will eventually ignite when the burner is repaired. Consider that if there was no safety shut down the ignition might have come on after there was oil built up inside of the boiler. This would cause a "puff back," an explosion, which is the worst possible thing that can happen to an oil burner. A major responsibility of all service technicians is to make certain that puff backs do not happen.

Resetting the primary safety switch has started the oil burner. You can see the mist of the oil spray inside the boiler but there is no spark to light it off. Shut the switch and remove the burner drawer assembly. Inspect the electrodes before removing them from their bracket. Is there carbon built up on the tips? Are any cracks showing in the porcelain? Is there crazing—a series of fine lines that resemble fingerprints—near the front end of the electrodes? Are the tips too close to the flame retention head or other metal parts of the oil burner? Are the bus bars or ignition wires in good condition? Any of these can cause no ignition and have to be checked out. There are no shortcuts in this process. Remove the electrodes from the mounting bracket and clean the porcelains. The porcelains must be white—shiny white if they are glazed—with no black lines that would indicate a crack. If there is the slightest doubt that the electrodes are not perfect, replace them.

Chart IV (Table 15-4) indicates that the nozzle should be replaced while the drawer assembly is out of the burner. Even though there was an oil spray, you may be eliminating another service call by checking the condition of the nozzle and the oil delivery system at this time. When the electrodes are installed in their bracket, be certain to adjust the setting of the tips to the burner manufacturer's specifications. If they are not available, follow these basic rules:

1. The tips must be 3/16″ apart.
2. The tips must be 1/2″ above the center of the nozzle.
3. The tips are always in front of the nozzle. The actual dimension is determined by the spray angle of the nozzle and the amount of air delivered by the burner. Narrow spray angle requires that the tips be farther forward than does a wide spray angle.
4. The tips must be at least 3/8″ away from the flame retention head or any other metal part of the burner.

The electrodes must be tight in the bracket. If the bracket is the type that has a set screw that directly holds the porcelain, you must place a shim between the screw and the porcelain.

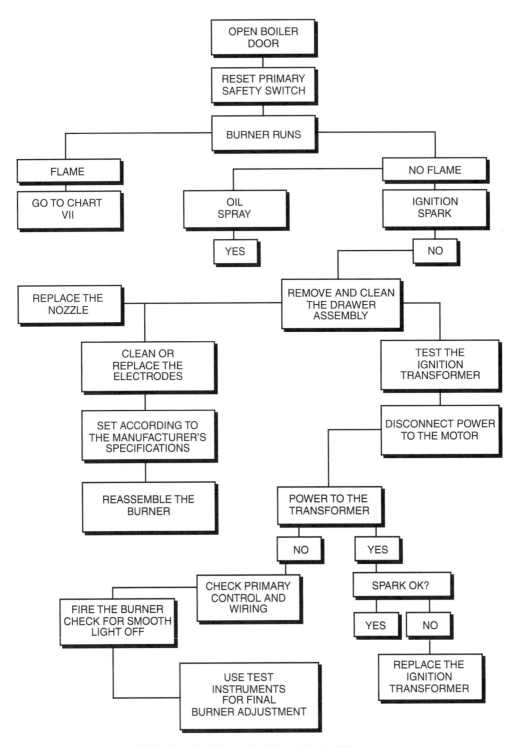

Table 15-4 **Chart IV. BNO—Locked Out on Safety—No Ignition.**

The aluminum straps used for pipe covering make an excellent shim for this purpose. Just cut off a small piece of the strap and slide it into the bracket between the porcelain and the interior of the bracket. This will allow you to tighten the set screw firmly without cracking the porcelain. The brackets that are split and do not have the set screw hold the porcelain directly may still require a shim to ensure that the electrode is held tightly in place.

Carefully check the high tension ignition wires. Be sure that the terminal clips are on securely and are actually making contact with the wire inside the heavy insulation. Check the insulation for cracks or signs of overheating or cuts from the burner fan. If there is any doubt about the condition of the high tension wires, replace them! Check the bus bars to make certain they are in solid contact with both the electrodes and the transformer high voltage terminals. Make sure they are not too close to any metal parts of the burner.

Before reassembling the burner, check the ignition transformer. Wipe the porcelain insulation around the high voltage terminals to get them clean and dry and to check for cracks or crazing. If the porcelain is cracked, the transformer must be replaced. You can check the transformer output by disconnecting the motor wiring so the burner does not run during the test. Open the transformer so you can see the high voltage terminals. Test the transformer with a high voltage tester designed for this purpose, by touching one of the terminals with the end of the tester while the tester lead wire is attached to the transformer case to ground it. A good transformer will give you a reading of approximately 5,000 volts on each terminal. Do not use a multimeter or voltmeter for this test because the high voltage will destroy them. If the transformer does not show the proper voltage, check the incoming voltage to the transformer with a regular voltmeter. If there is a reduced voltage, less than 120 volts at the primary coil, the secondary will not deliver the 10,000 volts. You will have to determine the cause of the low voltage condition and correct it.

Another method of checking the high voltage output of the transformer is by using a well-insulated screwdriver. Turn the power on and hold the blade against one terminal and close to the other terminal. There should be an active, intense spark generated. Slowly widen the gap and the spark should continue to at least 3/4″ without shutting off. Be extremely careful when working around an open transformer because of the possibility of a serious electrical shock. If there is the slightest doubt regarding the condition of the ignition transformer, replace it! If there is any indication that the ignition transformer was under water or got wet, it *must* be replaced! If burner shut down due to ignition failure is a continuous problem, you could try replacing the ignition transformer with an electronic spark generator. They are not susceptible to damp or low voltage problems and the hotter, more intense 14,000-volt spark may be able to vaporize and ignite contaminated or heavy fractions-laden fuel oil.

Before installing the drawer assembly, check the condition of the burner housing, fan, and air intake openings. It only takes a few minutes to clean the interior of the burner and this can eliminate a possible repeat service call. When the interior of the burner is clean, you can install the drawer assembly and close the burner to get it ready to fire.

Leave the motor wiring disconnected and turn the power on. Use a flame mirror to check the ignition. You should be able to see the spark directly above the nozzle in the exact spot for a smooth light off. This is a good time to check the safety switch lockout timing. How long does it take to lock out? 15 seconds? 30 seconds? 45 seconds? Longer? This is important because there was oil delivered into the boiler with no ignition that caused the original lockout. You should know how much oil is in the boiler because it will start to burn when you turn the repaired burner on and get a flame. Inspect the boiler fire box, flue gas passages, and flue breeching for signs of concentrations of liquid fuel oil. This can happen if the home owner has pushed the reset switch on the primary control several times in an attempt to start the burner before you arrived at the building. If you find oil in the boiler, use an oil absorber to soak it up and then vacuum the affected areas. In the rare case of a lot of oil in the boiler, you must contact the local fire department and request their assistance. Do not attempt to start the burner if there is a potential danger of a fire in the building. Most of the time you will be able to start the burner. Be prepared for some momentary smoke and vibration. Leave the boiler door open and stand off to the side. In a few minutes, the burner will be operating normally. Run the burner through several starting cycles to check the ignition repairs you have made. When the burner reaches steady operating temperatures, check the draft and smoke readings with the appropriate test instruments. Some service companies require an efficiency test after every burner repair. Efficiency testing and the use of test instruments are discussed in Chapter 19, Combustion Efficiency Testing.

USING CHART V. BNO—LOCKED OUT ON SAFETY— NO OIL: GRAVITY SYSTEM

Chart V (Table 15-5) and Chart VI (Table 15-6) deal with safety lockout caused by no oil delivery to the combustion process. The charts are separated by the type of oil tank installation, gravity feed, or two pipe. We will discuss the gravity feed systems first. In the majority of these systems, the tank is above the level of the oil burner and the oil runs by gravity to the fuel pump through a single pipe connected to the bottom of the oil tank. There is no suction required by the pump in this type of installation. There are single pipe installations where the burner is above the level of the oil tank. It is highly recommended that this type of installation be changed to a two pipe hook-up.

Resetting the safety switch has turned the burner on and you can clearly see that there is no oil spray into the combustion chamber. Find the oil tank and check the gauge to see if there is oil in the tank. If the tank is empty, immediately contact the oil dispatcher to find out if this is an automatic delivery account or if the customer calls for oil deliveries. The automatic delivery systems are very accurate and running out of oil may indicate a leaking tank or buried suction line. Check the bottom of the tank if it is accessible for signs of leakage. You can pressure test the suction line and tank, as outlined in Chapter 6, while waiting for the oil to be delivered. If the customer just forgot to order the oil there

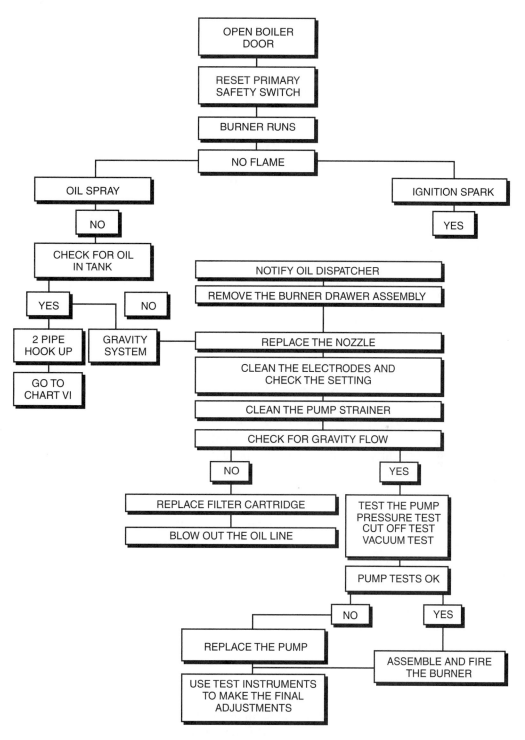

Table 15-5 **Chart V. BNO—Locked Out on Safety—No Oil: Gravity System.**

is nothing you can do until the delivery is made. Contact your office and let them know what you have discovered and ask for instructions. "Do I wait or do you have another service call for me?"

If there is oil in the tank, follow the chart. Start by removing the drawer assembly so you can service the nozzle. You can save a lot of time and do the customer a favor by replacing the nozzle. It is extremely difficult to clean a nozzle, particularly a small size one. Check the size, spray angle, and spray pattern before replacing the nozzle. Installing the wrong nozzle can really have a negative affect on the efficiency of the unit.

If the nozzle strainer is dirty, you can be certain that the pump strainer is equally dirty. Remove the pump strainer and check the gravity flow from the oil tank by placing a can under the open fuel pump and opening the valve on the oil line. The oil should run freely into the can, indicating a good gravity flow. If there is no oil or very little oil running through the oil line, you will have to check for an oil filter near the tank. If there is a filter at the tank, close the oil line valve and remove the filter cartridge. Place a can under the open filter and open the oil valve to test the flow of oil out of the tank. Reassemble the filter but do not install a new cartridge. Check for gravity flow at the burner again. If the flow is still slow, blow the oil line back with a CO_2 cartridge, or pump the line out with your hand pump. Remember, there must be a good gravity flow from the tank or the problems with the burner will not have been corrected. If the flow cannot be established, you will have to check the bottom of the tank for sludge or algae buildup. It may be necessary to have the tank professionally cleaned or you may want to install a floating oil line and switch to a two pipe hook up. Whatever you decide to do, it must be done before the burner can be reassembled and fired.

Once a good gravity flow is established, the filter cartridge can be installed. A new gasket must be used every time the filter is taken apart. The cartridge will not restrict the flow until it gets dirty, which should not happen for a long period of time unless there are problems with the tank such as sludge or algae buildup on the bottom. Water entering the tank will lead to these problems. It is essential that all connections on the fill and vent lines be tight so that no water can enter the tank through them. It is equally important to keep the oil tanks filled during the summer to prevent condensation buildup in the tank. Over a period of years, condensation can build up to where it can create serious problems inside the tank. When a new tank is installed, it is recommended that the tank pitch toward the outlet tapping so that small quantities of water will run through to the burner and not be allowed to accumulate inside the tank.

You can now install the cleaned or new pump strainer. Be sure to use a new cover gasket every time you remove the strainer cover. Open the oil valves and check the filter and strainer cover for signs of oil leaking. Now you can reassemble the drawer assembly. Do not forget to check the electrodes and the other components of the ignition system at this time. Install the drawer assembly close the burner and get ready to fire the burner. You will have to bleed the air out of the pump before the burner can fire. Install your pressure gauge in the pump if there is a separate port for the gauge, so that you can run a pressure test on the pump. If the gauge port is the same as the priming port, as on the Suntec J or Webster M pumps, you will have to prime the pump first and then install the gauge. Fire the burner and check the flame. Do the pressure test and check the flame. Run a cutoff test to check the pump pressure regulator seat. Run a vacuum test to check the pumping gears. If the pump fails any of these tests, replace the pump. Remove your gauge and run the burner through several starting cycles to check for smooth light off. Check for oil leaks and clean up around the oil burner. Use the test instruments, the draft gauge and smoke tester, to make the final adjustments. An efficiency test can be done when the unit reaches steady state temperature.

USING CHART VI. BNO—LOCKED OUT ON SAFETY— NO OIL: 2 PIPE SYSTEM

This chart (Table 15-6) deals with the problems relating to the two-pipe oil line installations. It would be a good idea to review Chapter 6 at this time so that you are familiar with how the oil is moved from the tank to the burner. Any questions? OK, reset the safety switch and see what happens. No oil spray? Is there oil in the tank? You may have to use a stick or ruler to measure the oil level in these tanks. Now would be a good time to check for water in the oil tank. Spread some water indicator paste on the bottom 6–8″ of the stick and drop it into the tank. No oil? Call the oil dispatcher and your dispatcher to find out what they want you to do. There is oil in the tank. Is there any water indicated? If there is now is the time to pump it out. Your truck should be equipped with a small pump and hoses for this purpose. Close the stick well cover tight and head for the boiler room where you can start working on the burner.

Remember, there are no shortcuts! Remove the drawer assembly and replace the nozzle. Check the electrodes and other components of the ignition system. Remove and clean the pump strainer. Be absolutely certain to use a new strainer cover gasket when you put the strainer back into the pump. Replace the filter cartridge if there is a filter on the job. Use a new cover gasket on the filter can. We are not concerned with oil leaking out, as on a gravity feed system, but with air leaking in and preventing the buildup of the required vacuum in the suction line. Check the interior of the burner and reinstall the drawer assembly. Close the burner and get ready to fire it. Install your vacuum gauge in the spare pump suction port or in a tee on the suction line between the pump and the oil line valve. Install your pressure gauge if there is a separate port for this purpose. You will have to bleed the air out of the pump and piping so you may have to install the pressure gauge after priming the pump. Open the oil valve and the pump bleeding valve and start the burner. Use a can under the bleeding valve to catch the oil as it comes out of the pump. Watch your vacuum gauge; it should be climbing which would indicate that there is suction building up in the line. A steady flow of oil should come out of the bleeding port. Once this flow is established, shut the burner off and install your pressure gauge. Now you can fire the burner. Check for smooth light off and allow the burner to operate while you run a pressure test on the pump. If the pump fails this test, it will have to be replaced.

The running vacuum on each installation is the result of the type of piping or tubing used, the distance the tank is from the burner, and the height difference between the

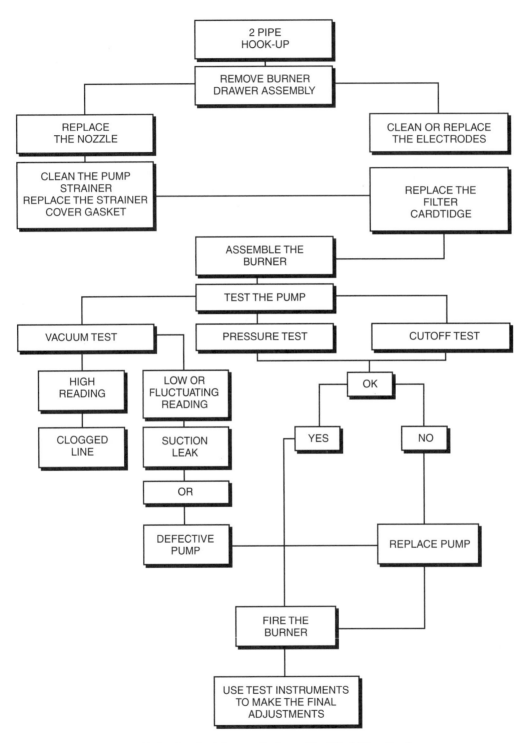

Table 15-6   **Chart VI. BNO—Locked Out on Safety—No Oil: 2 Pipe System.**

burner and the tank, which is referred to as the lift. A basic rule of thumb states that there is 1″ of vacuum required for every 10′ of horizontal run and 1″of vacuum required for every 1′ of vertical lift. Every new oil burner has an instruction sheet from the pump manufacturer as part of the instruction package. These sheets include a chart that indicates the maximum run and lift dimensions with 3/8″or 1/2″ tubing. The installer should design the piping to result in a running vacuum of 12″ or less. Higher running vacuums can result in vapor locks in the suction line, which can lead to safety lockout of the burner.

The pump must pass the vacuum test with no doubt in your mind that it is good. Once the pump is primed and your gauge is installed, close the suction line oil valve while the burner is running. A good pump will be indicated by the rapid movement of your vacuum gauge to 25″ or higher. Shut the burner off and watch your gauge. The vacuum should not drop off for at least 5 minutes. This will prove that there is no air leaking into the pump through the shaft seal, strainer gasket, or pressure regulator valve seat. Now you can test the suction line back to the tank. Open the oil valve and run the burner. You may have to prime the pump again. Check your gauge. Is there a good flow of oil to the burner at a steady vacuum? If there is, make note of the gauge reading because that is the normal operating vacuum for this particular installation. Many service companies have a service record card that is attached to the burner just for this purpose. Is your gauge fluctuating or bouncing up and down? That would indicate a suction leak that may require replacing the suction line. Some places that can be checked are all of the oil line fittings and the valve packing nuts. Only flared type connectors should be used on oil line piping to ensure airtight installations. If your gauge jumps to a high reading, a clogged suction line is indicated. You can attempt to clear the line by blowing it back to the tank with a CO_2 cartridge or by pumping the line with a hand-operated pump. Failure to clear the line would mean that the line has to be replaced. If the clog is caused by sludge buildup in the tank, either remove the sludge or install a floating suction line that will skim the oil from near the surface where the oil is clean.

Once all the tests are completed and the indicated repairs made, you can adjust the burner with the test instruments. Service calls in this particular category are often serious problems for service departments. By following the procedures outlined on this chart, you should be able to isolate and repair the problem on the first try. There can be no shortcuts taken or wild guesses made. Service technicians must be patient and methodical to become successful. Treat every service call as if you were working in your own home and your repeat calls will be kept to an acceptable minimum.

USING CHART VII. BNO—LOCKED OUT ON SAFETY— BURNER RUNS ON RESET

These service calls result in the most repeats. It is hard to resist just cleaning the stack switch helix or cad cell and leaving the job. That really is the wrong approach. To be safe you must assume that the burner tried to start and there was no ignition. Follow the procedure outlined in Chart IV first to check out the ignition system. Then go to the procedure in either Chart V or VI to check out the fuel delivery system. Did you find the problem? Now you can check out the flame detector (Step 3, Chart VII [Table 15-7]) on either the stack relay or cad cell.

Stack mounted primary controls lock out on safety if the pyrostat hot contact does not close. This can be caused by soot buildup on the helix, which acts as insulation and will not allow the helix to feel the heat of the flue gas. When you clean the helix, remember that it is a spring so check to see if it still has the required tension to operate properly. Either stretch it or turn it and see if it returns to its original position. If it does not, you may be able to replace the helix and solve the problem. Clean the pyrostat contacts with a contact spray and then polish them by rubbing a business card between them. Do not use a file on these contacts! Check to see that the hot contact opens when the pyrostat is in the starting or cold position. This is essential to ensure that the control will lock out if there is no flame. Run the burner and check the pyrostat action. If it is too slow, check the stack temperature to see if it is high enough to operate the pyrostat. Low stack temperature can be caused by under-firing the boiler, air leaking into the flue gas passages of the boiler, the stack switch being too close to the draft regulator, or the air adjustment to stack control being open too wide. Check the safety switch timing to see how long it takes to lock out. This can be accomplished by disconnecting the #3 wire and turning the burner on. The motor will not operate and the hot contact will remain open, causing a safety lockout. Most stack relays lock out in 70 seconds. If you have any doubts about the operation of the stack relay, change it.

Checking a cad cell requires the use of an ohmmeter. Disconnect the cad cell leads at the cad cell relay and connect them to your ohmmeter leads. With the flame off, the ohmmeter should indicate infinity or close to it. Turn the burner on and the ohmmeter should drop to 300 to 600 ohms. If it does not react in this manner, replace it. This is an excellent time to check the safety switch timing because the burner will lock out with the cad cell disconnected. Check the cad cell relay by connecting a 1,500-ohm resistor across the F-F terminals. This must be done by connecting one terminal, starting the burner, and then connecting the other terminal while the burner is operating. The relay should not lock out on safety. If it does it must be replaced.

Safety lockout may be caused by loose or dirty contacts in the low voltage circuit. If there is a problem of this type, the relay armature will attempt to close but because of reduced voltage to the coil, it will get hung up. Because the control thinks it is energized, and there is no flame, it will lock out. This should be the last procedure followed because if this is really the problem there is no possibility of puff back, which must be avoided at all times. That is why you must follow the procedures in the order shown on the chart so that any dangerous condition will be discovered and corrected. NO SHORTCUTS!

To Briefly Summarize

1. Servicing oil burners requires a systematic, logical approach.
2. Ask questions that could help to find out why the burner is not operating.
3. All BNO service calls must start at the primary control safety switch.

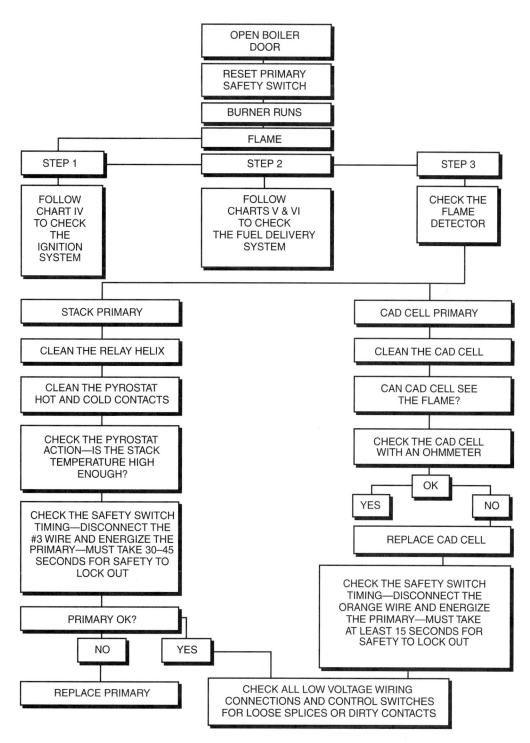

Table 15-7 **Chart VII. BNO—Locked Out on Safety—Burner Runs on Reset.**

4. Any time you find the primary safety switch locked out, you must remove the burner drawer assembly and thoroughly check the ignition system. This must be done to prevent the possibility of a puff back.
5. Use gauges and meters to check the equipment.
6. There are no shortcuts.

Please Answer These Questions

1. What are the factors involved in establishing the running vacuum for a two-pipe oil tank hook-up?
2. Explain the procedure for checking an ignition transformer.
3. Why does a stack mounted primary control lock out on safety?
4. Explain the procedure for locating a short circuit in the oil burner wiring.
5. Explain the procedure for testing a cad cell.
6. How is the safety switch timing of a primary control checked?
7. Why is the drawer assembly removed and cleaned on every safety lockout service call?
8. What are the reasons for motor overload switch trip out?
9. What are the reasons for water buildup in an oil tank?
10. What are sludge and algae?

CHAPTER 16

SERVICE PROCEDURES— IMPROPER OPERATION

Burner malfunctions do not always result in a complete shut down of the system. These service calls must be treated seriously because the homeowner has recognized that something is wrong and it is up to the service technician to make the necessary repair. The charts in this chapter deal with the variety of malfunctions common to residential oil burners. They all require the same systematic, methodical approach as did the BNO service calls. Listen carefully to the complaint and do your best to solve the problem on the first try.

Go to the boiler or furnace and do a visual inspection. Are you familiar with this installation? Did you do an annual tune-up here? Is this the first time you have seen this particular unit? Do you smell oil? Are there signs of oil leaking? Is there soot on top of the boiler or on the floor? Are all the boiler or furnace doors closed tight? Is the draft regulator in the flue pipe and is it functioning? Start the burner and check the flame. Is the combustion chamber in good condition? Is the flame too large or too small? Any of these can be the symptom of serious burner malfunction.

Use the chart that corresponds to the complaint as a guide to correct the problem. Remember, there are no shortcuts!

USING CHART VIII. IMPROPER OPERATION— PUFF BACK

A "puff back" is an explosion caused by delayed ignition of the air–oil mixture in the combustion chamber. The damage created depends on the size of the burner and how long it took for the mixture to ignite. Regardless of the severity, the puff back is the worst thing that can happen to an oil burner. Burners rarely puff without giving some warning such as starting vibration, oil smells, smoke, or soot around the boiler. For this reason the ignition system must be checked whenever the burner is serviced, even if there is no indication of ignition problems.

The service technician dispatched to a puff back service call must be prepared to face an extremely irate homeowner. A truly professional no nonsense approach is essential to placate the customer and make the necessary repairs. Follow the steps outlined in Chart VIII (Table 16-1). Start by cleaning the boiler room and boiler or furnace. You do not want to walk around on soot, which will probably be on the floor, because if you do it will soon be all over the house. While your vacuum is in the building clean the base of the chimney. Check the chimney for obstructions by holding your flame mirror in the base at an angle so you can see the sky through the top of the chimney. If you can't see the sky find out why and correct the condition if you can. Check the draft by crumpling a piece of newspaper into a ball, inserting it into the base of the chimney, and setting it on fire. A good chimney will draw all of the resulting flame, smoke, and eventually the burnt paper, straight up into the chimney. A blocked chimney will not act in this way and must be cleared to get the burner to operate properly.

The flue pipe, which connects the flue outlet of the boiler with the chimney, must be installed correctly and sealed. Horizontal runs of flue pipe should pitch up toward the chimney. The sections of pipe and the elbows should be held together with sheet metal screws. Replace any sections of the flue pipe that do not fit tightly together. The connection to the chimney must be sealed with furnace cement. Be careful not to have too much pipe inside the base of the chimney because this can partially block it. Install the draft regulator in a tee, which is part of the flue pipe. Always use the branch, or side outlet of the tee for this purpose. Set the regulator so that it is perfectly vertical and the damper is level so that it will operate properly. Do not seal or lock the damper closed because this will eliminate the regulator function (see Figure 16-1). On direct vent units (see Figure 16-2) you must check the intake and exhaust piping to make sure that they are not blocked. Repair any portions of the piping or exterior units that may have been damaged by the puff back.

Secure all the boiler or furnace doors or cleanout plates that may have blown open. This may require replacing some of the parts such as door hinges or latches. These parts may have to be ordered and temporary repairs are necessary to get the burner back into operation. Be sure that you follow up and make permanent repairs when the parts become available. Nothing looks worse than a wounded boiler with the doors held on with wire and furnace cement. Check the combustion chamber to make sure that all the bricks are in place and that there are no spaces where air can get in. Seal the chamber with refractory cement or by installing a wetpack-type ceramic liner if there are signs of damage.

Now you can work on the burner to find out why there was a puff back. Start with the drawer assembly because it is essential to be certain that the electrodes and bus bars or high tension wires are perfect. Replacing the nozzle is important, too because if the nozzle were to become partially clogged, the air–oil mixture might become too lean to ignite. This would allow the burner to build up the explosive mixture in the combustion chamber before ignition takes place, which will cause the explosion. If there is the slightest doubt regarding the condition of any parts of the drawer assembly, do not hesitate to replace them. The electrode setting must comply with the burner manufacturer's specifications, where available, or to the settings recommended in Chapter 7, The Ignition System.

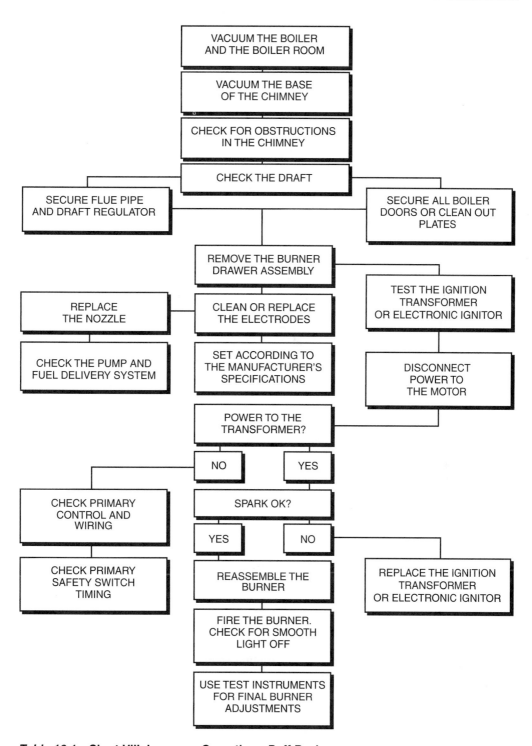

Table 16-1 **Chart VIII. Improper Operation—Puff Back.**

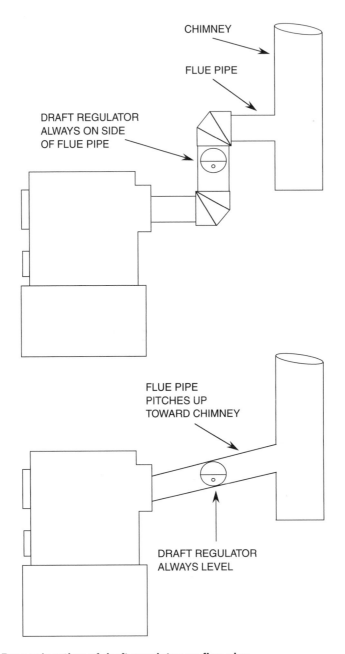

CHIMNEY

FLUE PIPE

DRAFT REGULATOR
ALWAYS ON SIDE
OF FLUE PIPE

FLUE PIPE
PITCHES UP
TOWARD CHIMNEY

DRAFT REGULATOR
ALWAYS LEVEL

Figure 16-1   **Proper location of draft regulator on flue pipe.**

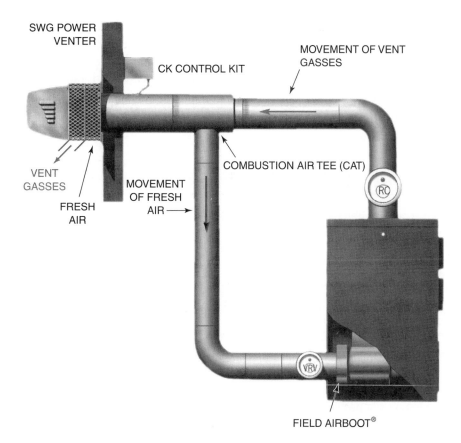

SWG POWER VENTER

CK CONTROL KIT

MOVEMENT OF VENT GASSES

VENT GASSES

MOVEMENT OF FRESH AIR

FRESH AIR

COMBUSTION AIR TEE (CAT)

FIELD AIRBOOT®

Figure 16-2 **A direct vent system can be severely damaged by a puff back and must be repaired before the burner can be put back in operation.** *Courtesy: Field Controls.*

The ignition transformer, or electronic ignitor, must be checked carefully. Open the transformer, or take it off the burner, so that you can inspect the porcelain insulation around the high voltage terminals. They must be smooth with no signs of cracks or crazing. Disconnect the power to the motor and turn the burner switch on so you can check the incoming voltage to the transformer. Reduced voltage conditions must be rectified so the full output of the transformer is available to create the ignition spark. This is an excellent time to check the primary control safety switch timing because all of this could have been avoided if the burner locked out on safety before the puff back occurred. Use a high voltage tester or a heavily insulated screwdriver to check the transformer. If there are any doubts about the condition of the transformer, do not hesitate to replace it!

Now is the time to check the fuel pump and the fuel delivery system. Clean the pump strainer if there was dirt in the nozzle strainer. Run the pressure, vacuum, and cut-off tests on the pump. If there are any doubts about the condition of the pump, replace it. Clean the interior of the burner and install the drawer assembly. Fire the burner and check for smooth light off. Allow the burner to run to heat up the chimney so that it will start to create the necessary draft. Check over the fire draft and adjust the draft regulator if necessary, to get −.02″ on your draft gauge. Adjust the air intake shutter; use a smoke tester to ensure that the burner is not creating smoke and soot. Run the burner through several starting cycles to make certain that it is igniting smoothly with no vibration or smoke. Clean up any residual soot and waste materials you have created. Be sure to make note of any followup work necessary on your service report.

USING CHART IX. IMPROPER OPERATION—VIBRATION

Vibration can occur when the burner starts or when it shuts off. Starting vibration must be treated as an ignition problem because it may very well be a symptom of a burner on its way to a puff back. Follow the procedure in Chart IX (Table 16-2), starting with the drawer assembly, and replace any parts that appear to be defective. There are no shortcuts.

On some large burner installations, delayed ignition and the resulting starting vibration can be caused by the motor's high ampere draw on starting. This can momentarily starve the ignition transformer for power, resulting in a delayed spark. This condition can be corrected by using a capacitor type motor that has a high starting torque. This will allow the motor to come up to speed quickly and reduce the length of time of the high ampere draw. Another solution is to install a delayed opening oil valve that will hold back the fuel delivery to the combustion process until the motor has come up to speed and there is no longer the problem of transformer power starvation. You might consider replacing the ignition transformer with an electronic ignitor. The starting vibration may be due to substandard oil that requires a hotter spark to vaporize and ignite. While you are at it, try a new primary control with pre- and post-purge cycles. If you do replace the primary control, use an oil solenoid valve that does not have a built-in time delay. Most of these primary controls have a 15-second trial for ignition. If you use a delayed opening valve, the trial for ignition would be cut to 12 seconds and that could lead to some nuisance safety lockouts. If cold oil is a problem, (is the oil tank outside?) you can install a nozzle line heater. As you can see, there are many things you can do. The choice is yours but the bottom line is to solve the problem!

Vibration may be caused by the chimney's inability to remove the flue gasses from the boiler. Check the base of the chimney and clean it if necessary. Check the operation of the chimney with a draft gauge or by burning a piece of newspaper in the base of the chimney. The draw of the chimney must supply the over the fire draft, −.02″, for the proper movement of flue gas through the boiler. Dirt or restrictive baffling in the flue gas passages of the boiler will have a similar affect. All the boiler doors and clean out plates must be secured so they do not rattle when the burner starts. The size of the nozzle and the pump pressure should also be checked because over firing the boiler may result in vibration.

Vibration at the end of the burner cycle can be caused by poor cutoff of the fuel by the pump pressure regulator piston. If the cutoff test, explained in Chapter 6, indicates

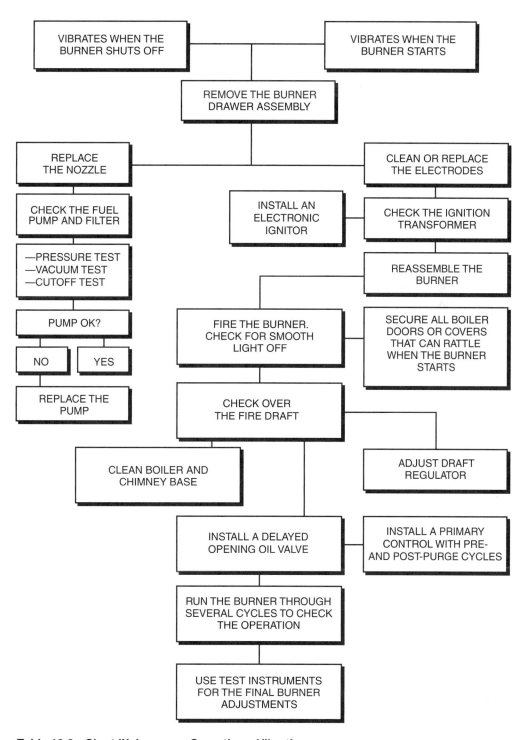

Table 16-2 **Chart IX. Improper Operation—Vibration.**

that this is a problem, you can either replace the pump or install a solenoid type oil valve. Remember that a delayed opening oil valve can solve starting vibration problems as well. These valves will cut the fuel off cleanly and solve the vibration problem at the end of the burner cycle. Another method of achieving clean cutoff at the end of the burner cycle is to use a nozzle with a spring loaded cutoff valve in the strainer.

USING CHART X. IMPROPER OPERATION— SMOKE AND SOOT

Black smoke and soot are the result of incomplete combustion because of inadequate air supplied to mix with the oil for the combustion process. White smoke is an indication that the air–oil mixture is too lean and that there is a problem with the fuel delivery to the combustion process. This will also result in a stinging odor that can be quite annoying. Very often the homeowner will complain that there is an oil odor in the building. In either situation, the condition must be corrected. Follow the steps outlined in Chart X (Table 16-3).

Black smoke indicates that there is a problem in the air delivery system. Yes, you have to start by removing the drawer assembly. Take a look at it. Is there lint built up on the stabilizer and electrodes? Very often the boiler room is used as a clothes drying area and lint does get into the burner. Lint built up on the drawer assembly is a definite indication that there is lint in the fan blades as well, and the fan as well as the air inlet, will have to be cleaned. Check the nozzle size to be certain that the boiler is not being over fired. Clean the interior of the burner and around the air intake shutter. Inspect the flue gas passages of the boiler and the base of the chimney for soot deposits. Restrictions in these spaces will affect the amount of air entering the combustion process. Use your vacuum and clean them out. Inspect the combustion chamber for broken or loose bricks that may be in the path of the flame. This will cause a carbon deposit generally known as a clinker, to form that will result in serious smoking of the burner flame. Check the fresh air intake into the boiler room to be certain that there is a continuous supply of oxygen for the combustion process. Once you have corrected any defects indicated in these procedures, you can reassemble the burner.

Fire the burner and run a pressure test on the pump. If the piston in the pressure regulator is stuck, you could get a high pressure reading that could lead to a smoky flame. While you are checking the pump, you might as well do a vacuum and cutoff test. If the pump is defective it will have to be replaced. If the pump tests OK, you can adjust the burner using the test instruments. Adjust the draft regulator so that the over the fire draft is −.02″. Adjust the air shutter on the burner so the smoke test indicates no more than a #2 smoke. The ideal setting is to have just a trace, #1, of smoke in the flue gas. Setting for 0 smoke may lead to the problem of white smoke and the odors that result.

White smoke contains water vapor and free oxygen that create a stinging sensation in the eyes and nose. The high quantity of water vapor will cause rusting of the flue pipe

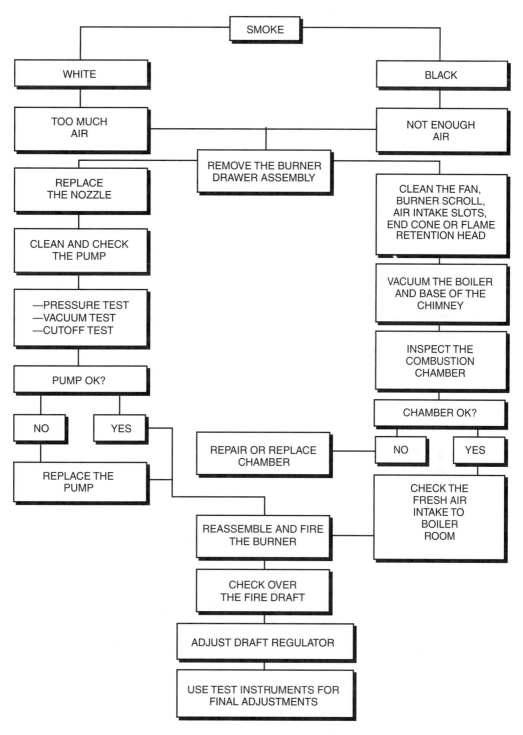

Table 16-3 **Chart X. Improper Operation—Smoke and Soot.**

and boiler parts. Running a burner with too much air in the combustible mixture will also reduce the operating efficiency of the unit. A partially clogged nozzle or a nozzle that is too small for the job could cause white smoke. Replacing the nozzle will eliminate this problem. The pump can be involved if the strainer is dirty or if the pressure regulator is defective and there is not enough pump pressure. If this is the problem, replace the pump. Air leaking into the combustion chamber can be a source of excess air and this must be corrected. Excessively high draft will suck too much air into the combustion process. This can be corrected by adjusting the draft regulator to get –.02″ of over the fire draft. Use the test instruments for the final burner adjustments.

USING CHART XI. IMPROPER OPERATION—OIL ODOR

Complaints of "oil odors" are quite common and are not always easy to satisfy. You have to locate the source of the odor, which may or may not be an oil leak. Chart XI (Table 16-4) identifies all the places where oil can leak and they all have to be checked. Check the bottom of the tank first to make sure that it is not wet. If there is a leak, you may be able to stop it with a magnetic patch until a permanent repair can be made. While you are at the tank, make sure that all the top tank tappings are closed securely, especially the plug used for measuring the oil in the tank with a ruler. Check the fill and vent piping for signs of leaks. Where is the vent cap? Is it close to a window? If it is, you may have to repipe the line to eliminate the problem. Check around the burner for leaks. Look inside the burner to see if it is wet. Perhaps the pump shaft seal is leaking. If it is, you will have to replace the pump. Use your flame mirror to check the nozzle and the front of the burner. Is there oil dripping out of the nozzle? Run a cutoff test on the pump to make certain that the pressure regulator is closing when the burner is off. Is the flame retention head wet? Is there oil inside the blast tube? These would indicate that there was a wrong angle or spray pattern nozzle in the burner and this would have to be corrected.

The problem of oil dripping out of the nozzle, usually referred to as "after drip," is serious and cannot be ignored. There are several possible causes of after drip. The most obvious is a defective pressure regulator in the fuel pump. If the regulator seat is damaged, the gravity flow of oil from the tank will drip through the nozzle. Check the pump with a cutoff test as described in Chapter 6. If the pump fails the test, it must be replaced. Another source of dripping oil is air trapped in the drawer assembly nozzle line. Older burners used to use a 1/4″ pipe for the nozzle line. Air inside the pipe would compress when the burner was running and then expand when the burner shut down. The expanding air would push a stream of oil out of the nozzle into the combustion chamber or burner blast tube. This problem can be eliminated by filling most of the inside of the pipe with a copper rod or pieces of copper wire. New oil burners have solved the problem by restricting the opening into the nozzle adaptor as well as making the interior of the nozzle line very small. The last major cause is reflected heat from the combustion chamber. Heating the oil in the nozzle causes it to expand and drip. The most susceptible units are

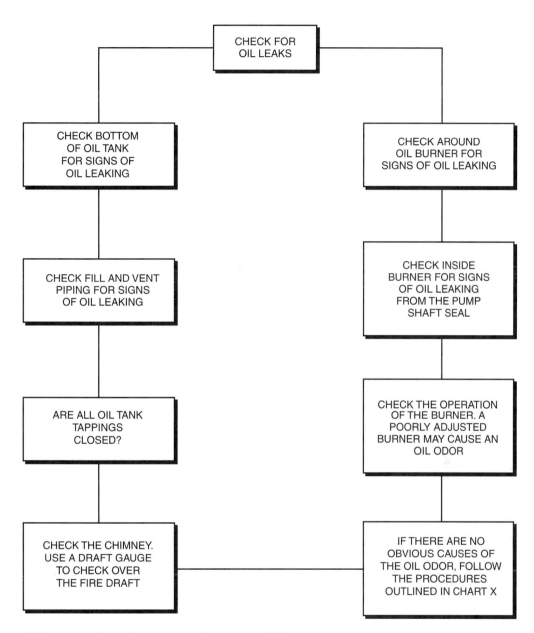

Table 16-4 **Chart XI. Improper Operation—Oil Odor.**

those with hard brick combustion chambers. The hard brick cools slowly and radiates heat for a long time. A possible solution is to install a Cerafelt liner in the chamber. This material cools quickly and may eliminate the after drip problem. Be sure that the burner blast tube pitches down slightly toward the combustion chamber and that the small drain hole in the end of the blast tube is open so that any oil that drips out will end up in the combustion chamber and not on the floor around the burner.

If you have located a leak and made the necessary repair, be sure to clean up any oil spots on the floor. This can be accomplished by using an oil absorbent compound, which is spread on the surface of the stain. Work the compound in until the stain and the odor are removed. Sweep up the compound and remove it and any other oily waste materials from the building.

Run the burner through a series of starting cycles to check for smooth light off. Open the boiler door and check the flame for odors. Adjust the draft regulator to get −.02″ of over the fire draft. Seal around the boiler doors, clean out plates, the flue pipe connection to the chimney, or any other place where flue gas might get into the boiler room. Be sure that there is a good supply of fresh air entering the boiler room.

USING CHART XII. IMPROPER OPERATION— NOT ENOUGH HEAT

Heating systems are designed to replace the heat lost from the building to the outside. As a service technician you are not involved in the design of the heating system. You do have the responsibility of supplying each system that you work on with a flame that contains the heat required for the system. Under firing any heating system, either through malfunctions in the fuel delivery system or installation of the wrong nozzle, will result in this type of service problem. Following Chart XII (Table 16-5), start with the burner to make sure that the flame is of the proper size. Check the boiler for either the nozzle size required or for a Btu input to determine the correct nozzle. Other methods of selecting the correct nozzle are outlined in Chapter 6. Use the test instruments to make the final burner adjustments. Run an efficiency test to determine how the boiler is reacting to the flame.

Each type of heating system has its own problems when it comes to not enough heat. The steam system controls must be set to allow the burner to run long enough to generate the steam necessary to fill all the radiation. The heat anticipator, a small heating element in the thermostat, which is normally set for the ampere rating of the primary control, will have to be set for a longer "on" cycle. The function of this device is to add a small amount of heat inside the thermostat cover so that the thermostat can reach the set point temperature and shut the burner off. Once the thermostat circuit is open, the heat anticipator cools off, allowing the thermostat to react to the room temperature and turn the burner back on when necessary. This results in a close control of the room air temperature. By setting the heat anticipator for a longer cycle, we can ensure that the thermostat will not shut the burner off until all of the radiation is full of steam.

The steam pressure control may get stuck in the open position for long periods of time. This can be caused by setting the "cut in" at 0 psi. The cut in must be set to at least 1 psi to prevent this. Another reason for pressure control "hang up" is dirt in the pig-tail siphon or boiler tapping. The pressure control must always be mounted above the boiler water level to make sure this does not happen.

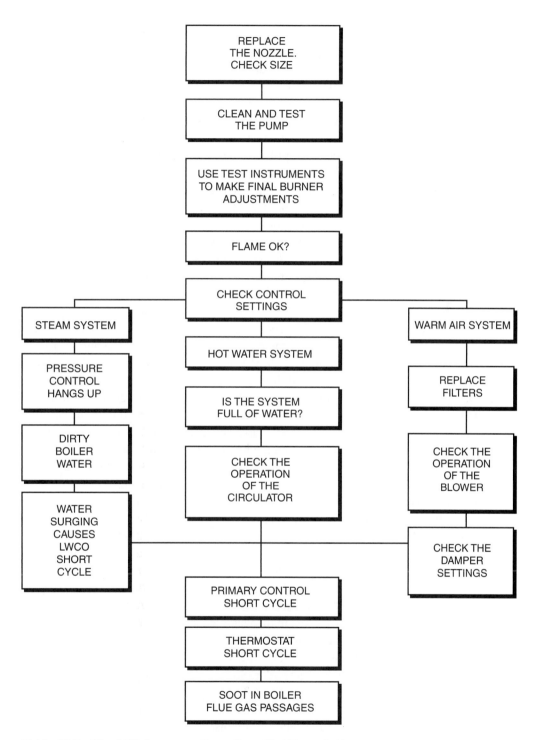

Table 16-5 **Chart XII. Improper Operation—Not Enough Heat.**

Dirt in the boiler water can lead to a "not enough heat" service call. Clean water boils at 212°F. Add some dirt and the boiling point is no longer 212° but something higher. How much higher depends on how much dirt is in the boiler. Where does the dirt come from? Most of the dirt is rust caused by free oxygen in aerated water supplied by community water supply systems. In addition to the rust, there may be some oil from the boiler installation pipe work and impurities in the local water. Not only will dirty water absorb heat and retard the steaming process, it can also lead to surging in the boiler. Surging of the boiler water, which is a rapid up and down movement, will cause the low water cutoff to shut the burner off unnecessarily. This short cycling of the burner will further impede the steaming process and result in a service call.

Flushing a steam boiler to remove built-up dirt is part of an ongoing maintenance program required for steam boilers. The building owner must be instructed to periodically remove a few pails of dirty water from the bottom of the boiler as well as the low water cutoff. Some steam boilers are equipped with a "skimmer" valve, installed at the water level of the boiler, which will skim any oil off the surface of the boiler water when it is opened. The skimmer valve should be piped toward the floor where a bucket can be placed under it to receive the hot boiler water.

If the boiler is not maintained properly it may become necessary to flush all of the water out to clean the boiler. Do not attempt to flush a hot boiler! Shut the burner off and leave it off for at least 2 hours before you start! When the boiler is cool enough to work on, connect a hose to the drain valve and drain the boiler water into a sewer. Your truck should be equipped with a small pump with a rubber impeller and hoses to enable you to drain the boiler quickly. Once the water is all drained, you can proceed with the flushing or washing process. Remove the steam safety valve and insert a hose in the opening. Wash the inside of the boiler with the water from the hose. Direct the nozzle to as many areas of the boiler as you can for the maximum effect. You will not get all the dirt out by doing this but you will get most of it! Reinstall the safety valve and fill the boiler using the normal boiler feed system. Drain it out again and then refill it. By this time, you can be reasonably certain that the boiler water is clean.

Now would be a good time to add a commercial brand of boiler cleaner to the boiler water. Use half the amount recommended in the instructions on the container. You can always add more but too much can lead to surging or odor problems. Never use boiler cleaning compounds on a dirty boiler without first flushing the water as described above. The reason for this is that the cleaning compound will precipitate the solid material in the boiler water, which can collect in the water jacket of the boiler and clog it completely. This can lead to overheating of the boiler sections and possibly to cracking the boiler. Check with the boiler manufacturer's literature or service representative before adding any cleaning compounds to their boiler. Some boilers have neoprene gaskets between the sections instead of steel push nipples. These gaskets may be adversely affected by the commercial cleaners and start to leak. One of the basic concepts of service work is to solve problems without creating other problems! Do not be ashamed to ask questions. Better safe than sorry.

Hot water systems have different problems that can lead to "not enough heat" service calls. Each system has a design temperature based on the amount of radiation in the building. The controls must be set to allow the burner to heat the water to this temperature. The high limit control setting determines how hot the water can get. It must be set at, or slightly above, the design temperature. If there is a low limit control in the circuit, it must be connected so that a call for heat will allow it to be overridden. Check the hot water system wiring diagrams in Chapter 14 to see how this can be accomplished.

A common problem with hot water systems is that portions of the system are not full of water. This can be the top floor radiation, which would indicate that the pressure reducing valve used to keep the system full is not set for high enough pressure. A general rule of thumb is that for each foot of altitude in the system, 1/2 psi of water pressure is required to keep the system full. If you cannot bleed the air out of the top floor radiators, you will have to adjust the pressure reducing valve. If you still cannot get enough pressure, you may have to replace the valve.

Air trapped in a baseboard radiation loop will result in that loop not heating. Check for traps in the piping, which may have to be removed. Check the diagrams in Chapter 9 to see how baseboard loops can be installed so that they can be completed vented. A particular problem is radiation that is installed below the supply main. These radiators and the piping connected to them may have to be filled with a hose and then connected to the piping. Air trapped in single-loop systems can be a serious service problem. Some contractors are now installing the circulators on the supply piping rather than the return line. This will result in pushing air in the piping through the system into the boiler where the air scoop can remove it. This is just one of the many variations you will encounter when servicing hot water systems so don't get upset if you see the circulators on top of the boiler instead of on the return lines at the bottom.

The circulators are generally not a source of problems but there is always the possibility of a loose coupling or impeller that can seriously affect the operation of the system. Unusual noise or water leaking can be signs of circulator failure. If there is a reverse acting control in the circulator circuit, it must be set at a temperature that will not create short cycling of the circulator operation. This control, which completes the circuit when the water temperature rises, should be set twenty degrees below the high limit control setting. This will ensure that the circulator will not short cycle and that the water pumped into the system is still hot enough to heat the building.

Warm air system controls have to be set so that only air warm enough to heat the building is sent into the ducts. The fan–limit switch must be set to accomplish this. Each unit must be set individually because this type of system is most sensitive to other conditions in the building such as insulation and orientation to the sun. There should be a differential of at least thirty degrees between the high limit setting and the fan switch setting to avoid short cycling of the blower motor. The thermostat heat anticipator should be set for a short cycle to avoid overheating the building.

Clogged air filters will certainly result in this type of service problem. Filters must be replaced once a year or more often if conditions require that this be done. Problems with the blower, such as a loose belt or dirt in the fan blades, must be corrected. The fan speed may have to be adjusted, if possible, to eliminate drafts caused by rapid air movement. Damper settings, particularly fresh air intake dampers, must be checked to be sure the air is circulating properly to heat the building. Check the individual supply and return registers to be sure they are not blocked by furniture or drapes.

Short cycling of primary controls or thermostats can cause problems with any heating system. Clean the open contacts in the stack relay or thermostat with contact spray or paper rubbed between the contacts. Remember, the spray is cold as it comes out of the can. This makes the thermostat close regardless of the room temperature. Wait a few minutes before you make any adjustments to the thermostat so you do not get it way out of calibration. Soot in the boiler will insulate the heat exchange between the flue gas and the heating medium and can result in not enough heat in the building. As you can see, there are overlapping problems involved in servicing oil burners. Experience will help you to identify potential problems that may not be related to the particular service call you are on. It is up to you to report the condition to the owner of the building or to your service department so that necessary repairs can be made before there is a service call.

USING CHART XIII. IMPROPER OPERATION— USING TOO MUCH OIL

As the price of fuel increases, homeowners become more concerned with the amount of oil they are consuming. This type of service call requires checking the delivery records (as indicated on Chart XIII [Table 16-6]) before a service technician is dispatched to the building. Fuel oil dealers can compare the amount of fuel used in a given period of time to a similar period of time by comparing the degree days of those two periods. Degree days are calculated by the difference between the average daily temperature and 65 degrees. For example, on a day when the average temperature was 48°, 17 degree days would result. The degree days are accumulated, generally starting on September 1. By the end of the month, the result could be 250 to 300 degree days. A customer might call and complain that last October he got a delivery of 175 gallons and this year his October delivery was 210 gallons. What happened? Maybe it was just colder this year, more degree days. If not, send a service technician to find out what is going on and correct the problem.

Were there any additions or alterations made to the building? People do strange things to their houses such as adding "sun rooms" on the north side of the building where the sun never shines. Did they finish the basement and install another heating zone to keep the pool table warm? If there were no changes or additions made to the building, you will have to check the burner to see why there was an increase in the amount of fuel consumed.

Start by removing the drawer assembly and replacing the nozzle. A fresh nozzle of the proper size, angle, and flame pattern, can do wonders for a heating system. Be certain to check the ignition system components while the drawer assembly is out of the burner.

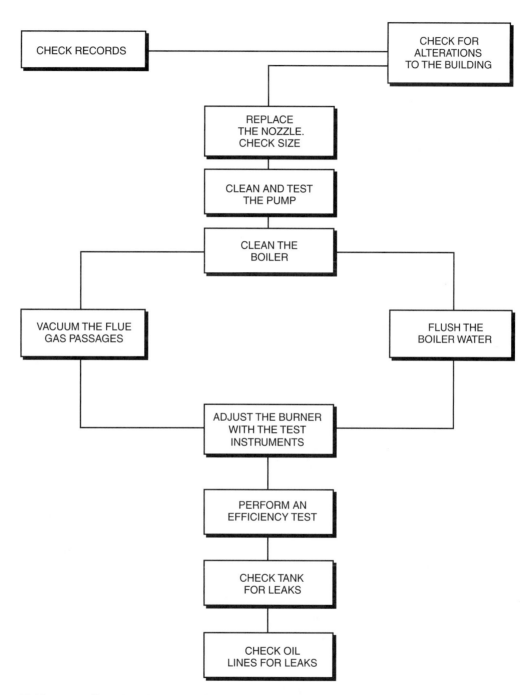

Table 16-6 **Chart XIII. Improper Operation—Using Too Much Oil.**

Clean the pump strainer and check for gravity flow, if appropriate, to see if the oil line filter is clogged. Replace it if there is no flow of oil from the tank to the burner. Reassemble the burner but do not fire it until you have checked the boiler flue gas passages for soot. An eighth of an inch of soot on the inside of the boiler can insulate the heat exchange between the flue gas and the water in the boiler, resulting in increased stack temperature and reduced boiler efficiency. This would be a good time to check the condition of the boiler water as well. Dirty water will raise the boiling point and could even form an insulating layer on the inside of the water jacket, which would have the same effect as soot in the flue gas passages.

Start the burner and test the pump to be certain that it is functioning properly. Adjust the burner with the testing instruments and run an efficiency test on the unit. Experience will teach you the kind of efficiency you can expect on the variety of heating equipment you have tested. New units should be capable of above 80% while an older unit will not get that high. Chapter 19, Combustion Efficiency Testing, deals with efficiency testing and methods of improving efficiency. Service records should include the results of previous efficiency testing on a particular unit so that you can find out if there has been a significant change.

Another problem you may face is a leak either in an oil line or a buried tank. Older buried tanks have been known to leak, which causes serious ecological problems. Many localities require testing of buried tanks. The company you work for, or will work for, may very well be involved in this type of testing. Pressure testing buried tanks and the piping connected to them requires that all of the tank piping, fill, vent, stick well, gauge line, suction, and return lines be disconnected and capped. A small air compressor can be used to build up pressure in the tank. A gauge can be installed on one of the piping lines, either the fill or vent, on the surface. The pressure required may be mandated by local laws. Thirty (30) psi should be enough to check for leaks in the tank and piping. Once the pressure reaches the required level, shut the compressor off and watch the gauge. If there are no leaks, the gauge will hold steady for at least five minutes. If the pressure on the gauge drops off quickly, you can be certain that there is a leak and the tank will have to be exposed to discover where the leak is. A soap solution spread on the piping joints will help locate leaking pipe work. If nothing shows up, you could have a bottom leak that would require draining the tank, opening it, cleaning the inside, and repairing the leak by painting layers of fiberglass on the inside.

A tank installation with a gravity feed could have a leaking suction line. This would certainly lead to an increase in the use of fuel. It is difficult to identify these leaks unless the pipe or copper tubing is on the surface and you can see the oil stain on the floor. Buried lines may be draining into sandy soil without a trace of a leak. These lines can be pressure tested, as outlined above, to determine if they are indeed leaking. Any questionable piping should be replaced.

By following these procedures, you should have discovered why there was an increase in fuel consumption. It is important to take these calls seriously and to make a real effort to discover the source of the problem and to correct it. The homeowner will certainly appreciate your efforts.

USING CHART XIV. IMPROPER OPERATION— BOILER LOSES WATER

This type of service problem is found mostly on steam heating systems and to a lesser degree on hot water systems. Loss of water can lead to burner shut down on boilers equipped with low water cutoffs, or to cracked boilers. You have to find out what is happening to the water by following Chart XIV (Table 16-7). Check around the outside of the boiler for water or signs of water on the floor. Boiler water will leave a rusty stain mark on a concrete floor. Look into the flue gas passages for signs of leaks. Water leaking into the boiler flue passes or combustion chamber will evaporate and not be noticeable from the outside of the boiler. Check the exposed piping around the boiler to see if there are signs of leaking. If you can, repair the leak.

Older steam boilers have been known to develop cracks above the water level of the boiler. This will allow steam to escape into the chimney during a burner cycle, leading to a loss of water due to evaporation. This can be detected by flooding the boiler. If there is such a leak, the water will run into the flue gas passages and you will see it. The solution to this problem is to replace the boiler! Water can also evaporate through air valves that do not close when steam enters them. Quick vents are serious offenders because of their size. A defective quick vent can allow enough steam to escape in a single day to shut the burner off. You can spot a defective air valve or quick vent when the system is in operation. They make a continuous hissing sound as steam escapes through the small opening in the valve. Any venting valve that does not shut off when steam enters it should be replaced.

Buried return piping must obviously be considered when looking for causes of water loss. Take into consideration the age of the system because the older it is the more likely there will be an underground leak. You have to prove to the homeowner, beyond any doubt, that there is a leak before starting an expensive repiping job. This should be the last place to look for the solution to this problem because repiping the returns first, without replacing defective air valves, will not really correct the original complaint. You may have a hard time collecting for the pipe work if the installation still loses water after you are finished. The method of identifying an underground leak requires shutting the unit off and disconnecting the return piping at the boiler. The boiler has to be drained, partially if there is a Hartford Loop connection, so it has to be allowed to cool off for a while before this work can be done. In any event, you want to be able to pour water into the vertical piece of pipe that comes up out of the floor from the buried piping. Fill it to the top and show this to the customer. Leave the job and do some other service calls or go to lunch. Go back two or three hours later and have the customer check the pipe with you. If the water is no longer at the top of the pipe, you can be quite certain that there is an underground leak (see Figure 16-3).

Hot water systems do not require the constant replacement of water that steam systems do. When there is a water loss problem in hot water heating, it is more often a loss

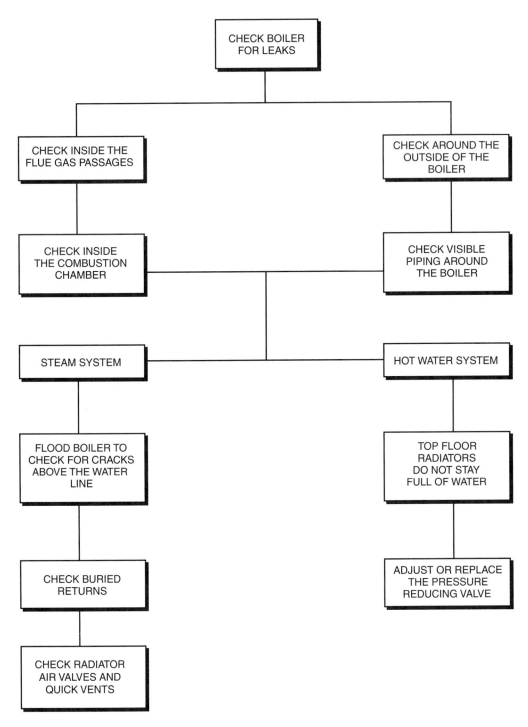

Table 16-7 **Chart XIV. Improper Operation—Boiler Loses Water.**

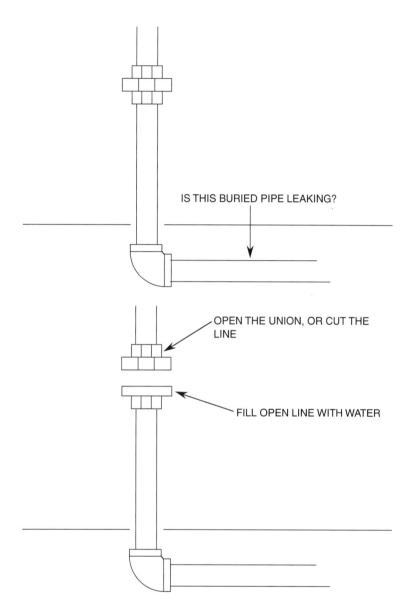

IS THIS BURIED PIPE LEAKING?

OPEN THE UNION, OR CUT THE LINE

FILL OPEN LINE WITH WATER

CHECK 2–3 HOURS LATER TO SEE IF LINE IS STILL FULL

Figure 16-3 **Procedure for detecting leaking underground piping.**

of the pressure required to keep the system full. Obviously, a leaking pipe or boiler would have to be repaired or replaced but there is no loss due to evaporation.

Keeping the system full is the responsibility of the pressure reducing valve, which can be either a separate unit or part of the dual valve. Pressure reducing valves reduce the incoming water pressure from the cold water main to the pressure required to keep the system full of water. The pressure required depends on the height, or altitude, of the system. Approximately 1/2 psi is required for every foot of altitude. Pressure reducing valves are set at the factory for 12 psi. This setting is adequate for a two-story building and would have to be adjusted for any additional system height. Once again the age factor is important. If the pressure reducing valve has been on the job for many years, it may have to be replaced. These valves have a strainer at the inlet port which can get clogged with rust or impurities in the water. Cleaning the strainer, if you can take it out after many years of service, may cure the problem. The labor and time spent in attempting to repair an old valve is pretty much the same as in replacing the valve and being certain that you have made the repair.

USING CHART XV. IMPROPER OPERATION— BOILER FLOODS: STEAM SYSTEMS

This chart (Table 16-8) is devoted to flooding problems encountered on steam heating systems. Make-up water for steam boilers requires some permanent connection between the boiler and the water piping in the building. This can be a simple hand feed valve, a mechanical water feeder, or an electrical valve controlled by the boiler low water cutoff switch. The failure by any of these valves to shut off completely will cause a flooded boiler. Flooding a steam boiler will not allow it to make steam to heat the building. There will be a lot of banging in the pipes and water squirting out of the air vents but no steam. If enough water gets into the system, its weight will force the steam pressure control to shut off the burner. If the boiler continues to flood, eventually the internal pressure will reach 15 psi, which will open the steam safety valve, releasing water into the basement. Sooner or later someone will discover the condition and call for help, so get your boots on and follow the chart.

The first step (not on the chart) is to drain the boiler to its proper level, which is about halfway in the gauge glass. If the gauge glass is dirty, clean or replace it so that both you and the homeowner can clearly see the water level. Check the hand feed valve first by opening the union directly below the valve. This union should be above the normal water level of the boiler so that when you open it, the boiler water will not back up out of the union. Does water drip out of the valve? Even a slow, continuous drip will eventually flood the boiler. If there is a drip, either replace the washer in the valve or the valve itself (see Figure 16-4). This operation may require shutting off the water main or other water valve in the building. Be extremely careful when closing old water valves, particularly gate valves, not to overtighten the valve. This could very easily break the gate off inside the valve and you will not be able to turn the water back on. It is better to leave an old valve slightly open and spill some water while working on a repair or replacement than to break the valve.

Mechanical water feeders can be checked in exactly the same way. The union on the discharge line, between the boiler and the valve, will probably be below the water level

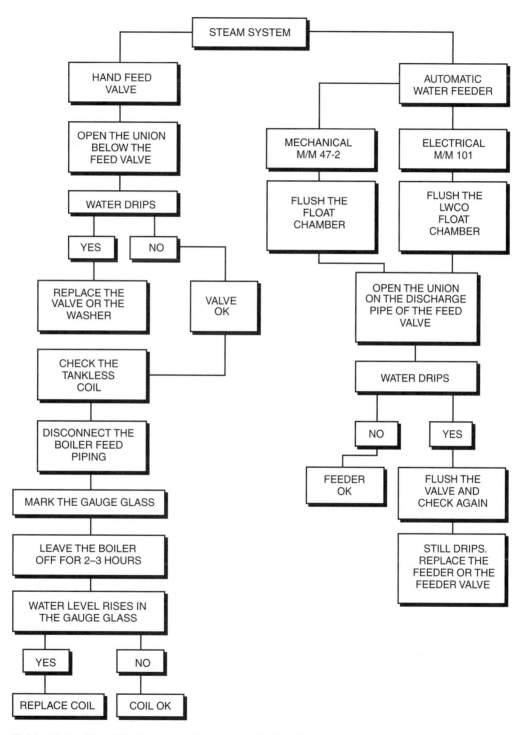

Table 16-8 **Chart XV. Improper Operation—Boiler Floods: Steam Systems.**

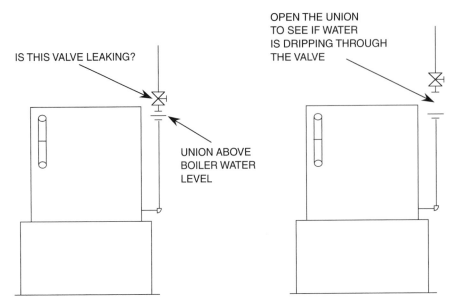

Figure 16-4 **Procedure for checking boiler feed valve when the boiler floods.**

of the boiler (see Figure 16-5). When you open this union, be prepared for the boiler water to back up out of the bottom pipe. Keep a bucket under the piping and force a rag into the open end of the union to keep the water from pouring out. Be careful because this water can be extremely hot. Flush the float chamber into the bucket and watch the action of the valve as water comes out of the pipe. Is the water clean? If it is rusty, you will have to flush the valve by moving the linkage assembly to completely open the valve. This should result in a good flow of water through the valve that will wash dirt or impurity buildup from the valve seat. Move the linkage to close the valve and see if water drips through it. If the water continues to drip, it is possible to replace the valve without replacing the entire unit. It does not pay to replace the valve if the entire unit is old. The float and the float chamber must be clean and the float must be able to move freely for these devices to operate properly. If water continues to drip through the valve or the float action is sluggish, replace the unit. Late models of the McDonnell Miller water feeders are equipped with replaceable cartridge type valves (see Figure 16-6). These units are identified by the hexagonal hub at the pipe connection ports on the valve. Replacing the cartridge is certainly easier than replacing either the mechanical or electrical water feeder.

Electrically operated water feed valves depend upon the switching action of the low water cutoff, either the float or the probe type, to know when to add water to the system. These units have to be thoroughly checked before performing the open union test on the water feed valve itself. Flushing the float chamber is a maintenance procedure that should be done periodically by the homeowner. This may not be happening at all so it is a good idea to flush low water cutoffs whenever you are in the building regardless of the service

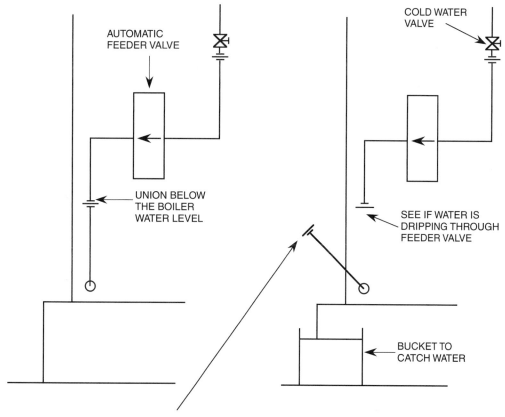

Figure 16-5 **Procedure for checking mechanical water feeder valve when the boiler floods.**

Figure 16-6 **Replaceable cartridge valve for McDonnell Miller water feeders.** *Courtesy: McDonnell Miller.*

problem. If the valves drip or the cutoff switches do not operate, they will have to be replaced.

If the tankless coil used to generate domestic hot water in the boiler leaks, it will flood the system. This is true for steam as well as hot water system boilers. You have to follow a procedure that will convince the homeowner that this is indeed the problem and that the coil will have to be replaced (see Figure 16-7). The only positive method of achieving this is to disconnect and plug the boiler feed system. Mark the gauge glass and leave the unit off for at least three hours. If the water level has risen in the gauge glass after that period of time, it definitely indicates a coil leak because there was absolutely no other way for water to get into the boiler.

Installation of a new steam boiler in an existing system can sometimes lead to flooding problems. These problems are not on the chart because they are rare, but you should be able to recognize them. The new boilers are much smaller and hold less water than their older counterparts. Old steam systems were assembled with large size piping and radiators. New boilers equipped with automatic water feed systems have a problem supplying enough steam to fill the piping and radiation without taking on additional water during a heating cycle. This would result in a flooded boiler when the cycle is completed and all the condensate returns to the boiler. The solution is to install a receiving tank and condensate pump system as the boiler water make up. This will allow the condensate to collect in a separate tank instead of returning to the boiler and flooding it. The water level control in the boiler operates the pump attached to the condensate tank when there is a need for boiler make-up water. Water make-up systems of this type are common to industrial high pressure boilers and steam generators (see Figure 16-8).

USING CHART XVI. IMPROPER OPERATION— RELIEF VALVE OPENS: HOT WATER SYSTEMS

Hot water systems can flood. This is indicated by the opening of the hot water relief valve on the boiler. The first thing that must be checked (as indicated in Chart XVI [Table 16-9]) is the expansion tank. The tank must be drained to create space for the air bubble, which can be compressed when the water is heated and expands. Water will absorb air so this must be done periodically on systems with plain expansion tanks. This requires some patience and common sense. First shut the valve between the expansion tank and the boiler. Then place a bucket under the tank drain valve and let the water out. There are special valves built to allow air to enter the tank through a long plastic tube so that the water will come out of the tank. If there is just a plain valve or hose cock, you may have to loosen a union to allow air to enter the tank. In any event, you have to remove a few buckets of water, depending on the size of the tank, to ensure that you have created the necessary space for the air bubble. The diaphragm type tanks have eliminated this procedure, that is until the diaphragm fails and then the tanks can fill completely with water.

Once you have the air space in the tank, just watch the pressure gauge for a while. If the pressure remains steady, turn the burner on and watch the gauge. It will rise but

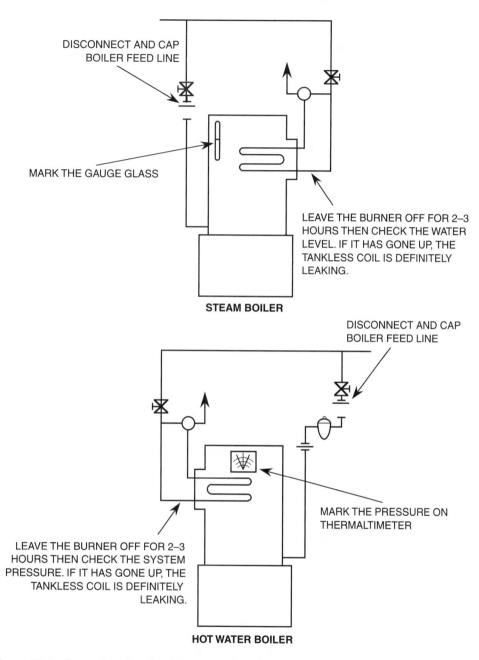

Figure 16-7 **Procedure for checking a tankless hot water heater when the boiler floods.**

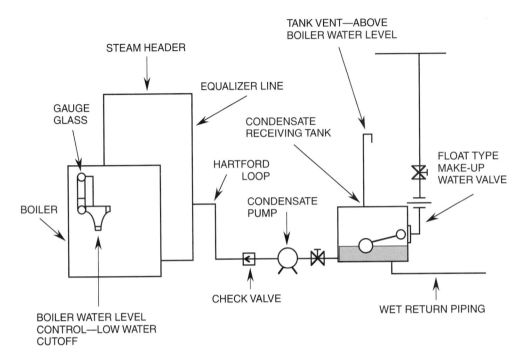

Figure 16-8 **Boiler feed system with condensate receiving tank avoids boiler flooding when new small boiler is installed as a replacement in an old steam heating system.**

remain below 30 psi, indicating normal operation of the system. You might want to check the top floor radiation to be certain that it is still full of water and vent it if necessary. If the gauge continues to rise with the burner off, try to adjust the pressure reducing valve. In the event that this does not stop the rising pressure, you may have to replace the valve. The other possibility is the tankless coil, if there is one in the boiler. If there is a tankless coil, disconnect the water feed line and plug it. Watch the gauge. If it continues to rise, the tankless coil is leaking and will have to be replaced. If the pressure remains constant, the pressure reducing valve is the culprit and that will have to be replaced.

To Briefly Summarize

1. Every type of service call requires the same systematic approach.
2. Pay attention to the homeowner's complaint and try to solve the problem on the first try.
3. Do not hesitate to replace burner ignition system parts if you suspect that they are defective. Puff backs must be avoided!
4. Develop methods or use those suggested, to prove to the homeowner that you have discovered the source of their problem before making an expensive repair.
5. Remember, there are no shortcuts!

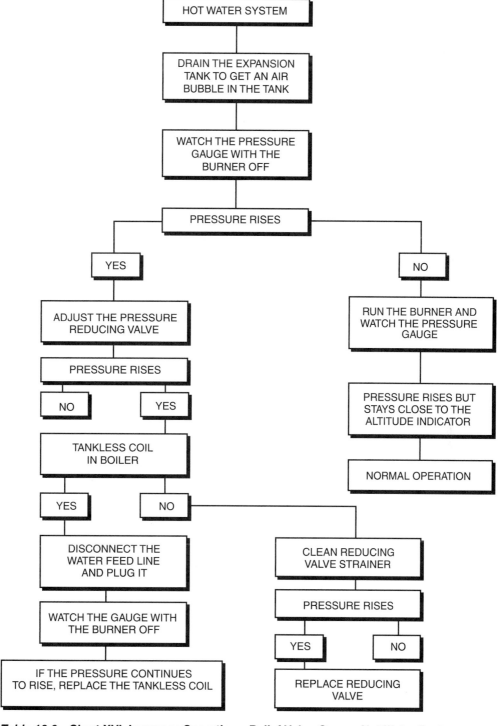

Table 16-9 **Chart XVI. Improper Operation—Relief Valve Opens: Hot Water Systems.**

Please Answer These Questions

1. Explain the procedure for checking the oil burner ignition system.
2. A steam boiler is losing water and you believe that a buried return pipe is leaking. Explain how you can prove that this is really happening.
3. A hot water heating system relief valve opens during the heating cycle. What is the most common cause of this problem and how can it be corrected?
4. Explain the procedure for pressure testing a buried fuel oil tank.
5. A steam boiler equipped with an automatic water feeder and a tankless coil is flooding. How can you tell which of these two parts is causing the problem?
6. What is the most common cause of "not enough heat" service calls for warm air heating systems?
7. How can vibration at the end of a burner cycle be eliminated?
8. Explain the relationship between draft and the oil burner smoking and creating soot.
9. Explain the function of the heat anticipator in the thermostat.
10. Explain how to deal with a customer's complaint that they are using too much oil.

CHAPTER
17 | DOMESTIC HOT WATER

The production of domestic hot water is an integral part of the residential heating system. This chapter is devoted to hot water and will describe the various methods employed to produce it. These methods are arranged chronologically and may or may not be used in your geographic location. The method depends on the type of heating system in the building. Steam and hot water systems are easily adapted to produce hot water while the warm air systems require a separate hot water generating system.

GENERATOR AND STORAGE TANK

The oldest method is the generator and storage tank connected as shown in Figure 17-1. The generator is a cast-iron casing that has tappings at the top and bottom so it can be connected to the boiler. This connection allows the hot boiler water to circulate through the case. Inside the case is a large-bore copper coil set vertically into the case and covered with hot boiler water. A circulation line is connected from the bottom of the storage tank to the bottom of the coil and from the top of the coil to the center of the tank. Cold water is fed into the bottom of the tank and the hot water leaves from the top of the tank. A pressure–temperature relief valve must be installed at the top of the tank.

When the boiler is in operation, hot boiler water circulates through the casing and heats the water inside the coil. As the water in the coil is heated, it rises into the storage tank. As the heated water enters the tank, colder water from the bottom of the tank moves into the generator coil where it is heated and rises into the tank. This gravity circulation will eventually heat all the water in the tank. Here are some problems with these systems:

1. There is a slow recovery rate. Once the hot water in the tank is used, it might take 6–10 hours to get another tank of hot water.
2. Dirt from the boiler water could collect inside the generator casing and insulate the coil. This requires flushing the casing and the boiler. The best method of keeping the

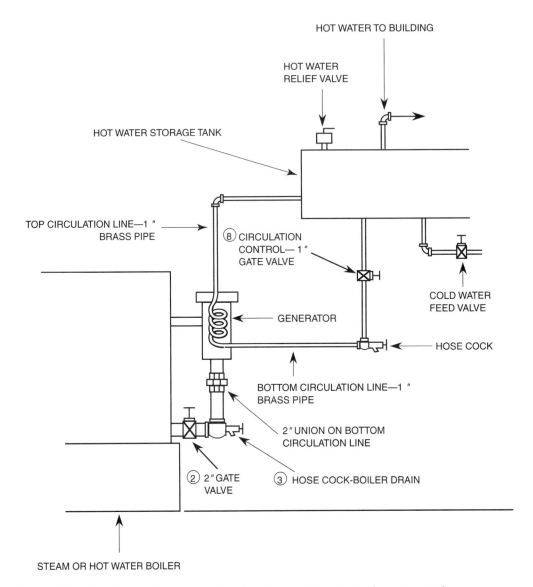

Figure 17-1 **Producing domestic hot water with a generator and storage tank.**

generator clean is to periodically flush water through it. This can best be accomplished by closing the valve (2) (Figure 17-1) first and then opening the drain valve (3). By doing this the boiler water will run through the generator case and clean it. Do not forget to open the valve (2) once the generator is clean. Opening the drain valve with the circulation valve open will keep the bottom of the boiler clean as well.

3. Sediment can build up inside the circulation lines between the tank and the generator. These lines have to be periodically flushed to keep them clean. By closing the

valve (8) on the bottom circulation line you can reverse flush the system, which really cleans it out.

4. Sediment and rust buildup in the bottom of the tank often leads to rusty water coming out of the hot water faucets. If you run the water long enough to clear out the rust, you may run out of hot water at the same time!

5. The hottest water is always at the top of the tank. Because of this, many tanks develop leaks at the top. Once this happens, the basement can get flooded because there is always water pressure inside the tank from the cold water line feeding it. There are several types of tank plugs made to stop this type of leak.

EXTERNAL TANKLESS HOT WATER HEATERS

The problems described above led to the development of the tankless hot water heater. The basic idea is the same as a generator except that the casing is larger and the coil is a small-bore, tightly wrapped unit that has a long path for the cold water to flow through. Figure 17-2 is a diagram of an external tankless coil connected to a steam boiler. The outside casing is piped exactly as was the generator. The coil is connected to the cold water at the bottom and to the hot water at the top. A temperature–pressure relief valve is installed in the cold water piping as close as possible to the coil. There is a valve on the cold water line to the coil that controls the hot water in the building. There is a mixing valve installed between the hot water coming out of the coil and the cold water entering the coil. The resulting mixed water then is piped to feed the hot water faucets in the building. Here are some problems with these systems:

1. Dirt from the boiler water collects inside the casing and insulates the coil. These units can be flushed in the same way as the generator casing.

2. Poor circulation of the boiler water through the casing will be more noticeable with the tankless coil than it is with a generator. It is possible to run cold water through the coil and cool off the water in the casing to the point that the water leaving the coil is cold. Poor circulation of boiler water through the casing will create this problem. To avoid this, a manifold was built on the boiler to supply water from all the boiler sections to the circulation line. The manifold is assembled by drilling and tapping the boiler sections, 6–12″ below the water level, depending on the boiler. A series of tees is connected to these tappings to form the manifold and the tankless coil casing is connected to the end of the manifold. The hot water control is installed in a tee on the manifold. Figure 17-2 is a diagram of this type of connection. Figure 17-3 shows a variation of the manifold known as a vacuumatic. In these systems, the casing of the coil is actually the manifold, which is connected to tappings made in the boiler as in the regular manifold installations. The casing is set above the boiler and connected to each section of the boiler. To fill the casing, the boiler must be flooded and the air vented out. Once the casing is full of water, the vent is plugged and the boiler drained to its proper level. The casing will remain full of water as long as no air can

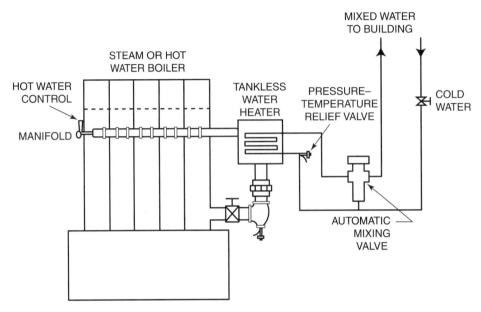

Figure 17-2   **Producing domestic hot water with an external tankless hot water heater.**

leak into it or any of the piping above the water level of the boiler. When the boiler is in operation, hot boiler water rises into the casing that heats the coil and the water inside it. A circulator and control can be installed to assist in the circulation of boiler water through the casing when necessary.

3. Tankless coils are rated in gallons per minute of cold water converted to hot water. Too much water due to high water pressure or a coil that is too small, will result in the water running cold after it is run for a period of time. This can be corrected by installing a flow valve that will restrict the flow of cold water through the coil to match the output rating.

4. The mixing valve may add too much cold water to the hot water coming out of the coil. An automatic mixing valve is necessary because the hot water coming directly out of the coil is too hot for use at a faucet. The mixing valve will also add to the volume of hot water available at the faucets because not all of the hot water generated has to pass through the coil on its way to the faucet. A properly designed system will take this into account. If there is a problem with the mixing valve, it may be possible to replace the thermostatic element in the valve. Restricting the cold water supply to the mixing valve will not solve problems created by the mixing valve element (see Figure 17-4). The piping to all automatic mixing valves must include a trap 12 to 18″ in length in the hot water line connection between the tankless coil and the hot inlet port of the mixing valve. Trapping this pipe will prevent overheating of the mixing valve element and prolong its life. This is true because during the periods when

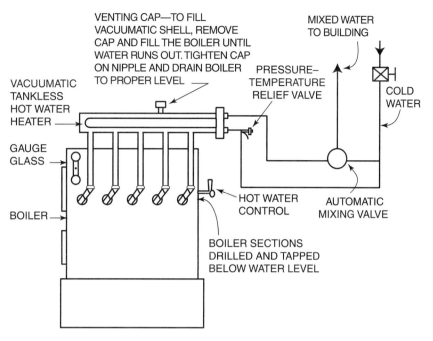

Figure 17-3 Piping connections for a vacuumatic type external hot water heater.

Figure 17-4 Automatic mixing valve is installed to maintain proper domestic hot water temperature. *Courtesy: Watts Regulator Company.*

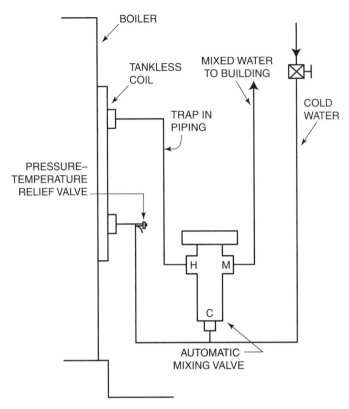

Figure 17-5 **Piping for automatic mixing valve must contain a trapped hot water line from the coil to keep the thermostatic element from overheating.**

no hot water is being drawn in the building, the hot water from the coil will not circulate down the trap to the element (see Figure 17-5).

INTERNAL TANKLESS HOT WATER HEATERS

Modern steam and hot water boilers are built to accommodate an internal tankless coil. These coils are attached to a plate that is bolted to the outside of the boiler (see Figure 17-6). The coil, which is made of finned copper tubing to allow a higher rate of heat absorption per running foot of coil, fits into the boiler below the water level. The additional load on the boiler created by the internal coil must be included in the boiler size calculations. Depending upon the number of people living in the building, the boiler size may have to be increased by 30–50% over the size required for heating. Failure to do this, particularly in a steam heating system, will result in the boiler's inability to raise pressure when someone is running the hot water. The water piping connections are identical to the

Figure 17-6 **Internal tankless hot water heater for Burnham V-1 boiler.** *Courtesy: Burnham Corporation.*

external tankless coil installations. It is possible to install two or more coils in a single boiler. The hook-up for this type of installation is shown in Figure 17-7.

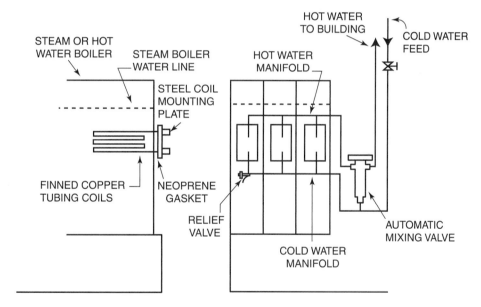

Figure 17-7 **Piping hook-up for multiple tankless coils in a large boiler.**

INDIRECT HOT WATER HEATERS

Indirect hot water heaters can provide hot water from either steam or hot water system boilers. These units are either the internal coil or external plate type heat exchangers. Figures 17-8 and 17-9 show internal heat exchanger units that employ plastic-lined, heavily insulated storage tanks. The piping connections for these units is shown in Figure 17-10. Figure 17-11 shows a flat plate heat exchanger unit that has a heavily insulated

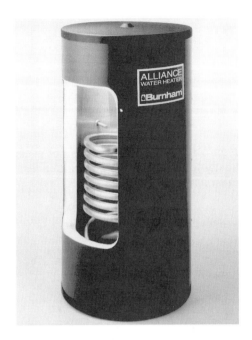

Figure 17-8 **Alliance indirect water heater has a stainless internal heat exchange coil.** *Courtesy: Burnham Corporation.*

Figure 17-9 **Indirect hot water heaters with internal heat exchange coils.** *Courtesy: Amtrol, Inc.*

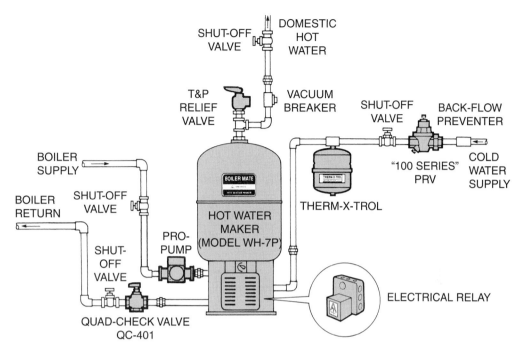

Figure 17-10 **Piping hook-up for Amtrol indirect hot water heaters.** *Courtesy: Amtrol, Inc.*

Figure 17-11 **Indirect hot water heater with external flat plate heat exchanger.** *Courtesy: Aero Environmental Limited.*

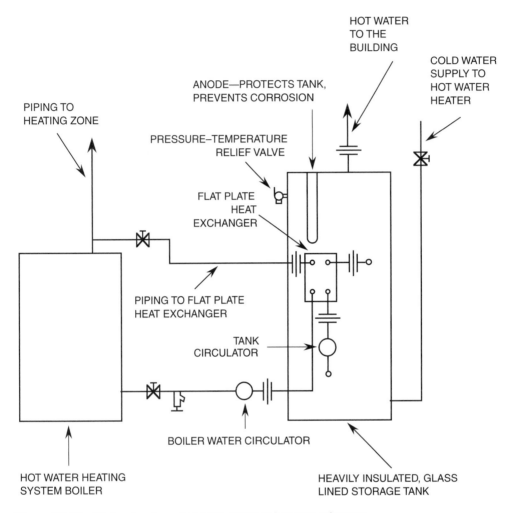

Figure 17-12   **Piping hook-up for Aero indirect hot water heater.**

glass lined storage tank. The piping hook-up for this type of hot water maker is shown in Figure 17-12.

   Boiler water from a steam or hot water boiler is circulated through the heat exchanger, which heats the water in the tank. The circulating pump is controlled by a hot water control in the storage tank. The major advantage to these water heaters is that they do not affect the boiler water temperature as directly as an internal tankless coil does. The ability to store hot water also helps to diminish the load placed on the boiler to produce domestic hot water.

OIL-FIRED HOT WATER HEATERS

The millions of homes that utilize warm air heating systems, as well as many steam and hot water systems, have separate hot water heaters. There have been attempts at adapting coils that fit into the furnace to the production of hot water. Most of these have been replaced with modern, well-insulated, glass-lined hot water heaters (see Figures 17-13 and 17-14). The fuel used for this purpose can be electricity, natural gas, propane, or fuel oil. The development of flame retention head oil burners, which can operate efficiently at low firing rates, has accelerated the use of oil-fired hot water heaters. They have a distinct advantage in recovery rate over the other fuels because of the intense heat of the oil flame. In many geographic areas, there may be a considerable saving in operating expense as well.

    The oil-fired hot water heaters are similar to small boilers. There is a fire box in the base of the unit where the fuel is burned. The fire box is lined with a combustion chamber

Figure 17-13 **Oil-fired hot water heater.** *Courtesy: Aero Environmental Limited.*

Figure 17-14 **Internal view of an oil-fired hot water heater.** *Courtesy: Bock Water Heaters, Inc.*

made of ceramic fiber or some other quick heating, insulating material. This is necessary because the short firing cycles involved in these units require quick heating combustion chambers to supply the proper environment for the total vaporization and combustion of the fuel oil. The insulating quality of the combustion chamber also protects the base of the unit.

The flue gas released by the flame travels through a center tube inside the storage tank to the flue gas outlet at the top of the heater. The center tube is 3–4″ in diameter, depending on the size of the water heater. Some manufacturers include a spiral baffle as part of the flue gas passage assembly. These baffles are removable for cleaning, which should be done at least once a year. This type of unit is susceptible to all the problems that regular oil-fired boilers have and must be treated accordingly. The over the fire draft and the flue gas connection to the chimney have to be correct to ensure the proper operation of the unit.

The storage tank itself is made of glass-lined heavy gauge steel. The glass lining ensures a supply of clean water flowing out of the hot water faucets as well as adding to the longevity of the unit. A magnesium anode, which is replaceable, is installed in the tank for additional corrosion protection. Each unit is equipped with a temperature–pressure relief valve to protect it. The cold water is supplied through a plastic dip tube that feeds the incoming water to the bottom of the tank where it is heated and rises to the top. The hot water is drawn from the top of the tank and sent to the faucets (see Figure 17-15).

The control circuit wiring includes a hot water control, either surface mounted or of the immersion well type, and a flame safeguard control. There should be a separate fuse and remote control switch, as well as a burner service switch, in the circuit. The oil line connection can be made to a tee on the suction line of a gravity feed tank hook-up. When the heating unit is connected to a two-pipe tank system, it is recommended that a separate suction line be used for the hot water heater. A common return line can be used for both burners if it does not result in elevated pump pressure on either burner. If the pressure is raised, the return line may have to be increased in size. Separate suction lines for each burner eliminates the necessity for the installation of check valves or isolation relays, which will allow only one burner to operate at a time.

Servicing oil-fired hot water heaters requires the same type of concentrated effort that working on the larger heating units does. Just because they are small is no reason to treat these units carelessly. "No hot water" service calls must be treated as BNO and the chart system should be followed. Improper operation service calls can be more difficult because the service technician must determine which burner is causing the problem, the heating or hot water unit, before the repair can be accomplished. An annual clean-up and the use of test instruments for final burner adjustments is highly recommended for oil-fired hot water heater installations.

The Toyotomi kerosene-fired water heaters can deliver 225 gallons of hot water an hour (see Figure 17-16). This unit is directly vented through the wall using the unique exhaust and combustion air intake system (see Figure 17-17). The heat source for this unit is actually a flame retention head oil burner. None of the parts, with the exception of the

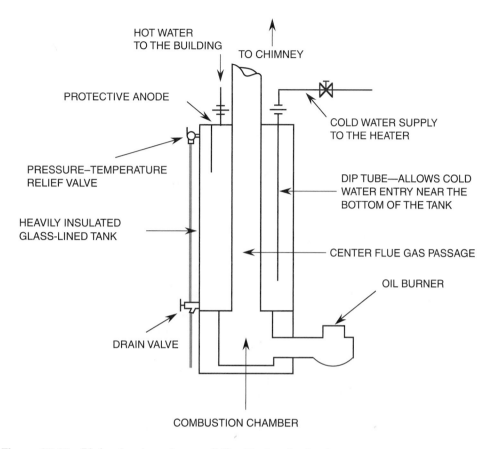

HOT WATER
TO THE BUILDING

TO CHIMNEY

PROTECTIVE ANODE

COLD WATER SUPPLY
TO THE HEATER

PRESSURE–TEMPERATURE
RELIEF VALVE

DIP TUBE—ALLOWS COLD
WATER ENTRY NEAR THE
BOTTOM OF THE TANK

HEAVILY INSULATED
GLASS-LINED TANK

CENTER FLUE GAS PASSAGE

OIL BURNER

DRAIN VALVE

COMBUSTION CHAMBER

Figure 17-15   **Piping hook-up for an oil-fired hot water heater.**

nozzle, are standard parts that can be found on any other oil burner. Figures 17-18 and 17-19 show the location of the components and safety controls.

The water is heated in the 5-gallon stainless steel heat exchanger. The thermistor's rapid response to temperature changes in the heat exchanger starts the burner as soon as water draw begins. Safety devices include a flame rod type flame detector, an air pressure switch to ensure air movement through the unit, an electrode type low water cutoff to prevent dry firing of the heat exchanger, and a fusible link in the oil supply line to shut the fuel off if there is a fire in the building near the heater. A standard pressure–temperature relief valve must be installed in the 3/4″ tapping that extends through the left side of the unit jacket. A tapping for a 1/2″ drain valve is also located on the left side of the heater.

Toyotomi water heaters are easily adaptable to be used as the heat source for radiant heating systems. Parts and training for service technicians are available through authorized Toyotomi distributors. It is highly recommended that service personnel attend these training sessions before they attempt to work on these water heaters.

Figure 17-16 **Toyotomi kerosene-fired direct vent hot water heater.** *Courtesy: Toyotomi U.S.A., Inc.*

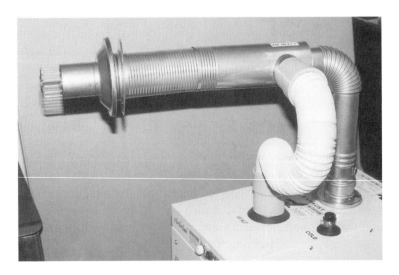

Figure 17-17 **Toyotomi direct exhaust and combustion air intake unit.** *Courtesy: Toyotomi U.S.A., Inc.*

Figure 17-18 **Component parts of the Toyotomi kerosene water heater.** *Courtesy: Toyotomi U.S.A., Inc.*

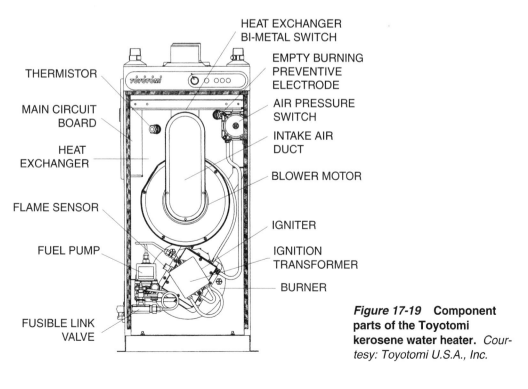

THERMISTOR

MAIN CIRCUIT
BOARD

HEAT
EXCHANGER

FLAME SENSOR

FUEL PUMP

FUSIBLE LINK
VALVE

HEAT EXCHANGER
BI-METAL SWITCH

EMPTY BURNING
PREVENTIVE
ELECTRODE

AIR PRESSURE
SWITCH

INTAKE AIR
DUCT

BLOWER MOTOR

IGNITER

IGNITION
TRANSFORMER

BURNER

Figure 17-19 **Component parts of the Toyotomi kerosene water heater.** *Courtesy: Toyotomi U.S.A., Inc.*

To Briefly Summarize

1. The production of domestic hot water is an integral part of all heating systems.
2. Steam and hot water systems can easily be adapted to produce hot water.
3. Warm air heating systems require separate hot water heaters.
4. As an oil burner service technician you will have to be able to repair any of the types of systems outlined in this chapter.

Please Answer These Questions

1. Explain how to flush the casing of an outside tankless coil.
2. Explain the operation of a vacuumatic tankless coil.
3. What are the functions of an automatic mixing valve?
4. Why is it necessary to install a pressure–temperature relief valve in a hot water storage tank?
5. What is the advantage to using finned type copper tubing for tankless coils?
6. Explain the operation of an indirect hot water heater.
7. Why is a dip tube used to feed cold water into an oil-fired hot water heater?
8. What is the function of the magnesium anode in a glass-lined water heater tank?
9. What type of combustion chamber is used in hot water heaters?
10. Draw a wiring diagram of an oil-fired hot water heater that includes the following:
    a. A fuse
    b. A remote control switch
    c. A burner service switch
    d. A high limit hot water control
    e. A low limit hot water control
    f. A cad cell flame safeguard primary control

ANNUAL TUNE-UP

There are some people in the oil burner industry who are promoting the idea that the annual tune-up of oil burners is no longer necessary. Nothing could be further from the truth. Certainly some of the new burners, installed with a direct vent and outside air intake system might be able to run for long periods of time without being serviced. But why take a chance on having a puff back or a no heat call? How about all of those older burners? They are still out there and need an annual treatment of tender love and care. This business is all about service. If you can't keep your customers warm and comfortable in their homes, you can be sure they will find someone who can.

The annual tune-up is not a welfare program for oil burner technicians. It should be a serious attempt to ensure the proper operation of the boiler-burner unit for the year. Under the best conditions, and if there are no mechanical break downs, a well-tuned oil burner will not require servicing from one annual tune-up to the next. Keep in mind the two factors involved in efficient operation: the combustion process as performed by the oil burner and the transfer of heat energy from the flame to the building heating system by the boiler or furnace. The tune-up of the burner is done to get it back to "good as new" condition so that it can completely burn the oil. Cleaning the boiler or furnace will remove the soot that insulates the heat exchange. Follow the procedures on Chart XVII (Table 18-1) and Chart XVIII (Table 18-2) to be certain not to miss anything. Remember, there are no shortcuts!

USING CHART XVII. CHIMNEY AND BOILER CLEANING

Is this really necessary? Until someone invents an oil burner that will not smoke and manufacture soot, it is not just necessary, it is essential. This phase of oil burner service work requires the same systematic approach as BNO or improper operation service calls. Annual tune-ups are usually performed during the summer when there is no demand for heat. Service departments make appointments with homeowners and ask that the units be shut down the night before the scheduled cleaning. Do not attempt to vacuum a hot boiler!

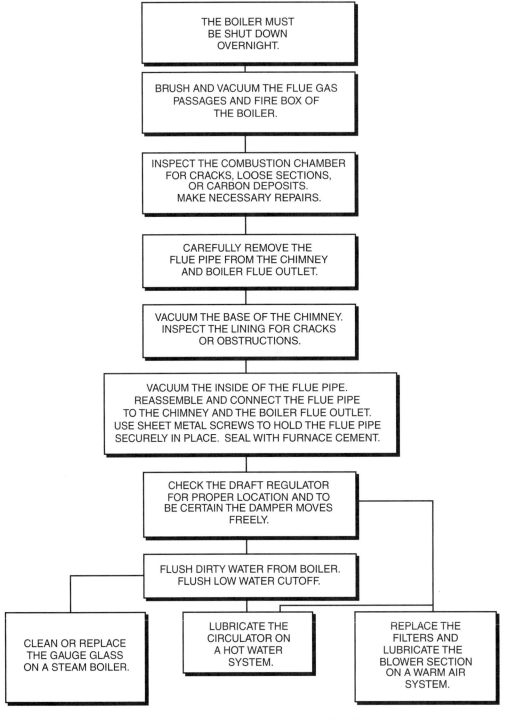

Table 18-1   **Chart XVII. The Annual Tune-up—Chimney and Boiler.**

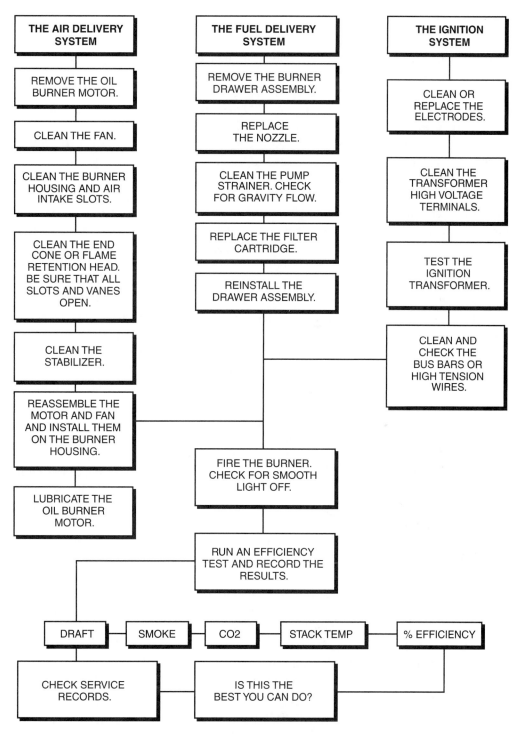

THE AIR DELIVERY SYSTEM

REMOVE THE OIL BURNER MOTOR.

CLEAN THE FAN.

CLEAN THE BURNER HOUSING AND AIR INTAKE SLOTS.

CLEAN THE END CONE OR FLAME RETENTION HEAD. BE SURE THAT ALL SLOTS AND VANES OPEN.

CLEAN THE STABILIZER.

REASSEMBLE THE MOTOR AND FAN AND INSTALL THEM ON THE BURNER HOUSING.

LUBRICATE THE OIL BURNER MOTOR.

THE FUEL DELIVERY SYSTEM

REMOVE THE BURNER DRAWER ASSEMBLY.

REPLACE THE NOZZLE.

CLEAN THE PUMP STRAINER. CHECK FOR GRAVITY FLOW.

REPLACE THE FILTER CARTRIDGE.

REINSTALL THE DRAWER ASSEMBLY.

THE IGNITION SYSTEM

CLEAN OR REPLACE THE ELECTRODES.

CLEAN THE TRANSFORMER HIGH VOLTAGE TERMINALS.

TEST THE IGNITION TRANSFORMER.

CLEAN AND CHECK THE BUS BARS OR HIGH TENSION WIRES.

FIRE THE BURNER. CHECK FOR SMOOTH LIGHT OFF.

RUN AN EFFICIENCY TEST AND RECORD THE RESULTS.

DRAFT — SMOKE — CO2 — STACK TEMP — % EFFICIENCY

CHECK SERVICE RECORDS.

IS THIS THE BEST YOU CAN DO?

Table 18-2   **Chart XVIII. The Annual Tune-up—The Oil Burner.**

Hot soot can set the paper filter bag in the vacuum on fire and create an unbelievable mess! Another potential problem is trying to vacuum a boiler that has been completely shut down for a long period of time. Condensation inside the flue gas passages turns the soot into a sticky mess that will get all over you and whatever you touch. If you run into one of these beauties, you will have to run the burner for a while to dry out the inside of the boiler. Try not to get it too hot or you will have problems handling hot soot.

Start by brushing the flue gas passages of the boiler or furnace. The access openings may be a door or plates that have to be removed. Start from the top and work toward the combustion chamber where all of the soot collected can be picked up by your vacuum cleaner. Use a soft wire brush and try to get the inside of the unit as clean as when it was new. There may be baffle plates or spiral baffles inside the fire tube boilers. These have to be removed, cleaned, and reinstalled. Do not discard the baffles! There may be scale deposits inside the boiler. These are caused by condensation and will have to be scraped off because they act as insulation between the hot flue gas and the heat exchange surface of the boiler. Check the interior of the boiler for signs of leaks around the tubes or between the boiler sections.

Inspect the combustion chamber for cracks or broken sections. You may have to report back to your dispatcher if repairs are necessary so that he can contact the owner and set a price for the work that has to be done. Now is the perfect time to install a ceramic fiber liner in the chamber. This will not only seal the chamber, it will help the fuel vaporize and burn completely because of its ability to reflect heat back into the flame almost as soon as the burner starts. Small cracks in the chamber can be repaired with retort or furnace cement. Try to avoid covering large areas with either of these products since they will peel and may cause an impingement problem if the oil spray hits the peeling area and starts to form a carbon deposit. Brush the inside walls of the furnace area and vacuum all the soot and loose pieces of refractory material out of the combustion chamber.

Carefully remove the flue pipe that connects the boiler flue gas outlet to the chimney. This may require removing sheet metal screws that hold the sections of pipe together. Brush and vacuum the inside of the flue pipe. Vacuum the base of the chimney. You may have to break up some hard material that has been deposited in the chimney base by rain coming down the chimney. Place a mirror in the base of the chimney and adjust its position until you can see blue sky at the top of the chimney. If you cannot see the sky, there may be an offset in the chimney that will have to be cleaned. This can be done by dropping a burlap bag with some weights in it down the chimney on a rope. A weighted brush can be used as well. You can check the chimney by crumpling a piece of newspaper up and placing it into the opening at the base of the chimney. Light the paper with a match and if the chimney is clear, the flames and smoke from the burning paper will be drawn up into the chimney. Once the chimney is clean, you can reassemble the flue pipe. Use the screws to hold it together and pay particular attention to the operation of the draft regulator. Be sure it is set level and that the damper gate swings freely. Seal around the base of the chimney and any other boiler access openings with furnace cement.

Drain some water out of the bottom of the boiler and the low water cutoff. This is necessary to remove sediment that acts as insulation on the inside of the boiler. Clean or replace the gauge glass on a steam boiler. You may have to drain and flush the boiler to get the waterside clean. Now is the time to do this because the boiler is cold and you should have the time to spend. Hot water system boilers should be checked for sediment as well as steam boilers. If the unit has a probe-type low water cutoff, this is the perfect time to remove and clean the probe. This is particularly true if you are working on a steam system boiler. Check the inside of the boiler at the probe tapping to be sure there is no rust or sludge buildup that could possibly ground the probe. A grounded probe will signal the control that the water level is okay when in fact it might not be so, and that could lead to a dry fired and eventually cracked boiler.

Lubricate the circulator(s) in a hot water system. Replace the filters in a warm air system. Vacuum the inside of the blower section and as much of the plenum as you can reach. Lubricate the blower section and check the belt to see if the belt is cracking and if it has the proper tension. Secure all the boiler doors and clean out plates. Vacuum around the boiler and boiler room to clean up any soot that may be on the floor. Wipe the outside of the boiler or furnace jacket with a clean rag. Now you can start on the oil burner.

USING CHART XVIII. THE ANNUAL OIL BURNER TUNE-UP

Each of the burner systems must be worked on during the annual tune-up. Remember that you are trying to get the burner back to its original, brand new condition. Start by removing the burner drawer assembly. Set it aside and work on the air delivery system by removing the fan for cleaning. While the fan and motor are off the burner housing, it is a simple matter to clean the inside of the burner. Pay particular attention to the air intake slots and the air adjustment shutter. Oil burners "breathe" the surrounding air that can contain lint, dust, animal hairs, and soot. Any or all of the above end up inside the burner and in the fan blades. Be certain to check the front end of the burner, either the air cone or flame retention head, to clean any accumulated carbon deposits that could restrict the air flow coming out of the burner. Check the front of the burner with a mirror to make sure it is clean. Flame retention heads that are attached to the blast tube present a serious problem because they do get dirty and are hard to clean. You have to clean all the slots and vanes to get the full affect of the retention head. Since the boiler is still cold, you can stick your hand into the combustion chamber and reach down to the head to clean it. In some cases, the burner may have to be removed to get the head cleaned. Burners that have the retention head attached to the drawer assembly are much easier to service. Clean the fan blades and the face of the motor. Once the housing and fan are clean, you can reassemble the motor and fan on the burner housing. Lubricate the oil burner motor with a few drops of #10 motor oil in each oil cup.

Work on the drawer assembly next. Remove the nozzle and electrodes and clean the exterior parts of the assembly. Run some water through the oil pipe to clean it. Now is the perfect time to install a new nozzle. If your company keeps accurate service records, you should have been given the correct nozzle for each burner you will be working on before you left for work. Your service vehicle should be equipped with a supply of nozzles that are best kept in their plastic containers in a steel case. Check the nozzle size, spray angle, and spray pattern when replacing nozzles. The wrong nozzle can reduce the efficiency of the unit.

Clean the electrodes and check for signs of cracks or crazing. Cracks in the porcelain appear as black lines while crazing looks like a fingerprint or rough area made of fine black lines. Either cracks or crazing will allow the high voltage electricity from the transformer to leak through the porcelain insulator and short out the spark. Unglazed porcelains that cannot be cleaned or glazed porcelains that show signs of crazing or cracks MUST be replaced! Install the electrodes in their brackets and set them according to the burner manufacturers specifications or by eye. Remember, the oil spray must not hit the electrode tips because this will build up a carbon deposit that will eventually short-circuit the spark. The idea is to have the spark just graze the edge of the air–oil mixture as it leaves the end of the burner. This will ensure a clean, smooth light off on every heating cycle. Clean and check the bus bars or high tension wires. Make certain that these parts fit securely onto the end of the electrode and come into solid contact with the transformer high voltage terminals. Clean the transformer terminal porcelain and check for cracks or crazing. This would be a good time to check the transformer as well. Once all of this is completed, you can reinstall the burner drawer assembly.

The pump is next, so remove the strainer and check for gravity flow of oil if this is a single-pipe system. Replace the filter cartridge as well. Be sure to replace the strainer cover and filter can gaskets so that there are no leaks. Fire the burner and check for oil leaks and smooth light off. You may decide to test the pump if you think there is a problem with the flame. Check for leaks around the pump outlets and the nozzle line connections. Run the burner and adjust it by eye. Let it run for a few minutes to warm up the chimney. You can use this time to clean up any debris you may have created during the clean-up and remove it from the building. Now you can find out how effective all of this work has been by running the efficiency test.

Before you leave the job, shut the burner off and take a few minutes to check the inside of the blast tube for signs of liquid fuel. Liquid fuel can build up inside the blast tube if the nozzle spray pattern is incorrect for the burner, or the distance between the nozzle and a fixed retention head is not set properly. If there is oil in the blast tube, you must clean it out and correct the problem to avoid another service call for an oil smell or oil leaking around the burner. To carry this scenario to its extreme, the result of oil in the blast tube could very well be a fire, and you do not want to be the last service technician to have worked on it.

To Briefly Summarize

1. The annual tune-up is an important part of the oil burner technician's work and must be taken seriously.
2. The idea is to have the burner operate from one annual tune-up to the next without breaking down.

3. The boiler and chimney cleaning are done to restore the unit to as close to the original condition as possible.
4. The oil burner should get a new nozzle as well as any other parts that are questionable.

Please Answer These Questions

1. Why does the boiler have to be cold before it can be cleaned?
2. What effect does scale inside the flue gas passages have on the unit operation?
3. Explain how you clean the boiler.
4. Explain how you can check a chimney.
5. Why is it a good idea to install a fiber refractory blanket in the combustion chamber?
6. How do you seal the base of the chimney where the flue pipe is attached?
7. Why does sediment have to be removed from the bottom of the boiler?

CHAPTER 19

COMBUSTION EFFICIENCY TESTING

One of the most important responsibilities of the oil burner technician is to ensure that the equipment is operating at its maximum efficiency. Every time an oil burner is serviced, the instruments used in the efficiency testing should be used to make the final burner adjustments. This is the only way to know if each individual unit is operating at maximum efficiency. There are many variables in determining the maximum efficiency of a boiler–burner unit. The experience of doing many tests on a variety of heating systems will help you understand the problems involved and what you can expect to be able to accomplish on each installation.

The word efficiency means performing a task quickly, precisely, and with a minimum of energy expended. An efficiency percentage can be determined by using the following formula:

$$\% \text{ Efficiency} = \frac{\text{Energy Output}}{\text{Energy Input}}$$

When the efficiency of a heating unit is to be determined, the energy input is the result of the oil burner burning the fuel to release the heat energy locked in it. The energy output is the amount of heat delivered to the building heating system by the boiler or furnace. The test must include a method of determining how well the boiler is accepting the heat energy during the process. The best time to test the efficiency of a heating unit is when it is new and being fired for the first time. The results of this test should be recorded and left on the job as part of a service record (see Figure 19-1). Many local building codes require an acceptable minimum efficiency rating for approval of the installation. Once the original rating is recorded, it can be used as a benchmark to check the operation of the unit as it gets older. A thorough test should be done when the annual tune-up of the burner is done.

The test procedures described below are performed with the Bacharach Fyrite Testing Kit (see Figure 19-2). There are more modern, electronic kits available as shown in Figures 19-3 and 19-4. Your company may equip their service vans with any one of these test kits. The Fyrite kit is discussed here because it is most common because of its low cost and rugged nature.

Figure 19-1 Service record on boiler indicates results of previous efficiency tests.

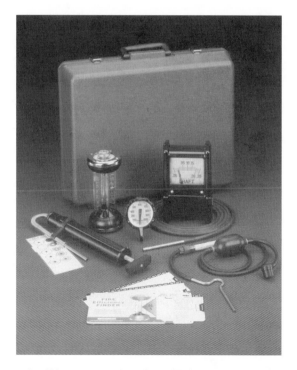

Figure 19-2 Bacharach oil burner combustion efficiency kit. *Courtesy: Bacharach, Inc.*

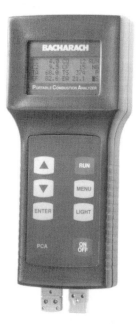

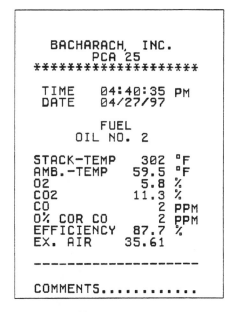

```
      BACHARACH, INC.
           PCA 25
********************

   TIME     04:40:35 PM
   DATE     04/27/97

          FUEL
        OIL NO. 2

STACK-TEMP      302  °F
AMB.-TEMP      59.5  °F
O2              5.8  %
CO2            11.3  %
CO                2  PPM
0% COR CO         2  PPM
EFFICIENCY     87.7  %
EX. AIR       35.61

--------------------

COMMENTS.............
```

Figure 19-3 **Portable combustion analyzer provides a printout of flue gas test results, which includes carbon monoxide (CO).** *Courtesy: Bacharach, Inc.*

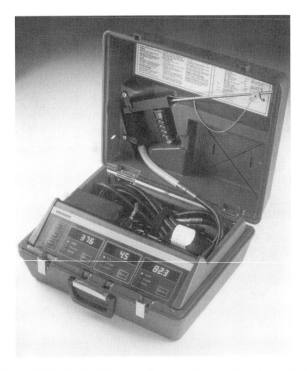

Figure 19-4 **Bacharach Model 300 Combustion Analyzer.** *Courtesy: Bacharach, Inc.*

The instruments included in the test kit are a draft gauge, a smoke tester and smoke chart, a CO_2 tester, and a stack thermometer. It is necessary to prepare the test instruments before the test begins. Place the draft gauge (see Figure 19-5) on a flat surface and set it at the 0 mark (see Figure 19-6). Unwind the rubber sampling tube and install an extension piece in the end so an accurate over-the-fire draft reading can be obtained. Install a filter paper in the smoke tester (see Figure 19-7). Have the smoke chart handy so you can use it quickly. Prepare the CO_2 tester by "wetting" it (see Figure 19-8). Turn the unit upside down and allow the liquid to be spread inside both ends of the unit. Stand the tester up and allow all of the liquid to drain into the bottom. Push the valve at the top of the tester down to vent the unit and allow all of the liquid to drain into the bottom reservoir. Set the scale at the 0 mark (see Figure 19-9). If the liquid does not reach the mark and the scale cannot be adjusted to get to the level of the liquid, you will have to add a little water. Be careful to add only a few drops so you do not flood the tester. Put the sampling unit close at hand so you do not have to search for it. Do the same with the stack thermometer so the entire test can be done quickly (see Figure 19-10).

It may be necessary to prepare the unit for testing by drilling a 1/4″ hole in the front door of the boiler and another 1/4″ hole in the flue pipe approximately 12″ from the boiler flue outlet. Do not drill this opening too close to or after the draft regulator. These openings can be sealed by pushing a 1/4″ stove bolt into them when the test procedures

Figure 19-5 **Bacharach draft gauge.** *Courtesy: Bacharach, Inc.*

Figure 19-6 **The draft gauge must be set level and at the zero mark before the over-the-fire-draft reading can be taken.** *Courtesy: Bacharach, Inc.*

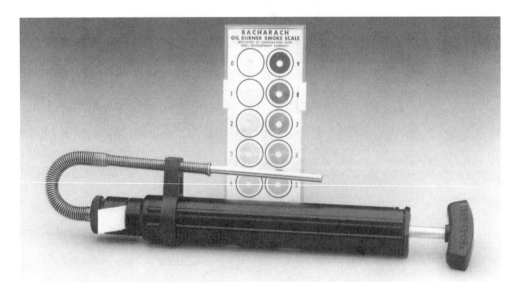

Figure 19-7 **The Sure Spot Smoke Tester and smoke chart.** *Courtesy: Bacharach, Inc.*

Figure 19-8 **Preparing the Fyrite Flue Gas Analyzer for a test by wetting it thoroughly.** *Courtesy: Bacharach, Inc.*

Figure 19-9 **The Fyrite Flue Gas Analyzer ready for a test.** *Courtesy: Bacharach, Inc.*

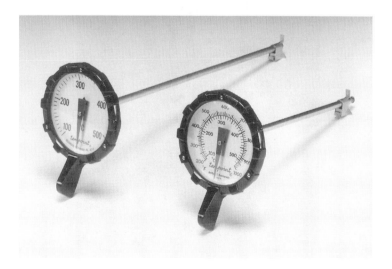

Figure 19-10 **Stack thermometers.** *Courtesy: Bacharach, Inc.*

have been completed. The opening in the boiler door is where the over the fire draft is checked. The opening in the flue pipe is where the smoke test, CO_2 test, and the stack temperature are taken. Turn the burner on and allow it to operate for a few minutes to warm up the chimney so the testing procedure can begin.

THE DRAFT TEST

The draft test is done first because it is necessary to have the over the fire draft set at −.02″ so the burner can supply the air necessary to complete the combustion process. This test is not necessary if the unit is a new pressurized boiler with a forced draft oil burner. Set the draft gauge down on top of the boiler and set it at the 0 mark. The gauge should be standing level so it does not have to be held during this test. Push the draft sampling tube into the boiler through the hole in the boiler door. Adjust the draft regulator so the gauge indicates an over the fire draft of −.02″ (see Figure 19-11). This can easily be accomplished with most residential oil burners. Problems with draft and solutions will be discussed later on in Chapter 20, Improving Combustion Efficiency.

Figure 19-11 **The draft gauge, during the over-the-fire-draft test, with the draft regulator set so the draft is −.02″ of water column.** *Courtesy: Bacharach, Inc.*

THE SMOKE TEST

The smoke test is next, so take a sample of the flue gas with the smoke tester and compare the resulting spot on the filter paper with the smoke chart (see Figure 19-12). To obtain a sample of the flue gas insert a piece of filter paper in the tester and lock it in place. Then insert the end of the sampling tube through the hole in the flue pipe. Be careful that the end of the sampling tube does not come into contact with the inside of the flue pipe because soot may get into the tube and ruin the test. Pump the tester 10 times to get a sample of flue gas to pass through the filter paper. Remove the paper and compare the spot on it with the smoke chart in the test kit. Adjust the air shutter so that the result of the smoke test is a #2 smoke or less (see Figure 19-13). Any smoke reading higher than #2 will result in the formation of soot inside the boiler. The soot will insulate the fireside of the heat exchanger and cause the stack temperature to climb and the boiler efficiency to drop. Opening the air shutter too much may result in a wet spot when the smoke test is done. This is the result of too much water vapor in the flue gas caused by the excess air entering the combustion process. It is best to leave the burner operating with just a trace or #1 smoke.

Figure 19-12 **Taking a flue gas sample during the smoke test.** *Courtesy: Bacharach, Inc.*

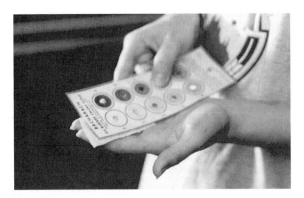

Figure 19-13 **Comparing the spot on the filter paper with the smoke chart.** *Courtesy: Bacharach, Inc.*

THE CO$_2$ TEST

The CO$_2$ test is done as soon as the smoke test is completed (see Figure 19-14). Insert the sampling tube in the hole in the flue pipe and squeeze the bulb five or six times to clear the air out of the tubing and the filter. Press the sampling tube firmly into the valve on the top of the tester to open the valve. Squeeze the bulb 18 times to collect the maximum amount of flue gas the unit can hold. You cannot overfill the unit. Once the sample has been collected, quickly remove the sampling tube from the top of the tester and turn the tester over to allow the liquid to absorb the CO$_2$ in the sample. Stand the tester up and allow all the liquid to settle in the bottom reservoir. A slight vacuum is created inside the tester by the absorption of the CO$_2$ by the KOH (potassium hydroxide) solution. This vacuum will cause the diaphragm in the bottom of the tester to push the liquid into the center tube of the tester. The result is that you can read the percentage of CO$_2$ in the flue gas by how high the liquid rises in the center tube (see Figure 19-15). The perfect combustion process will yield 15.3% CO$_2$ in the flue gas. If it were possible to attain this percentage on an oil burner, it would indicate that there was no excess air in the combustion process. As the percentage of CO$_2$ in the flue gas decreases, the amount of excess air increases. Excess air traveling through the flue gas passages does not contribute to the heat exchange. In fact, the excess air will raise the stack temperature indicating a greater loss of heat through the chimney into the air. This will definitely reduce the operating efficiency of the unit. You might think from the information above that every burner is adjusted to get to the magical 15.3% CO$_2$, but that is not the case. The problem is that the oil burner breathes the surrounding air and over a period of time, the parts of the air delivery system do collect lint and dust (see Figures 19-16 and 19-17). The accumulated material will cut down on the amount of air delivered to the combustion process and this will lead to a smoking oil burner. It is common practice to open the air shutter when the CO$_2$ percentage is 13 or 14%, to drop it down to 12%. The additional, or excess, air will have a slight

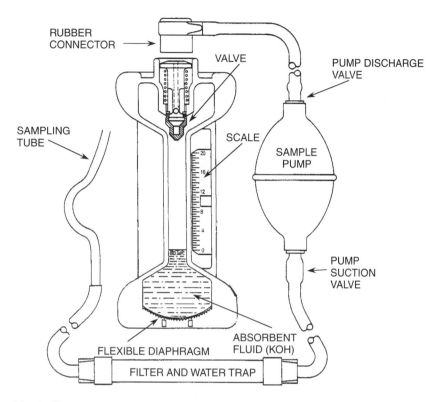

Figure 19-14   **The parts of a Bacharach Fyrite Flue Gas Analyzer.** *Courtesy: R. W. Beckett Corp.*

Figure 19-15   **After collecting the flue gas in the analyzer, the fluid rises to indicate the percentage of CO_2 in the flue gas.** *Courtesy: Bacharach, Inc.*

Figure 19-16 **Willie likes the boiler room. His hair ends up in the oil burner fan and air handling parts.**

Figure 19-17 **Lint from the laundry equipment in the boiler room ends up in the burner air delivery system.**

effect on the overall efficiency of the unit but in the long run it will help to prevent soot buildup in the boiler or furnace.

THE STACK TEMPERATURE TEST

Insert the stack thermometer in the opening in the flue pipe and allow the burner to operate (see Figure 19-18). The net stack temperature is the highest point reached by the thermometer less the room air temperature. As an example, if the stack thermometer reaches 520°F and remains steady, subtract the room temperature—70°F—to determine the net stack temperature, which is 450°F (see Figure 19-19). Stack temperatures below 350°F should be avoided on conventional boilers since this can lead to damage in the flue breeching and flue pipe because of condensation. However, new high efficiency units may operate at lower stack temperatures. These units are equipped with plastic flue breechings that have drains or drip pockets where the condensation can be disposed of without damaging the flue pipe.

Determining the Combustion Efficiency

When the tests are completed, the burner can be shut off and the efficiency percentage can be found by using the chart contained in the efficiency testing kit (see Figure 19-20). Set the slide at the stack temperature and move it down to the CO_2 percentage. The efficiency percentage is indicated in the slide window. Compare it with the service record to see how well you have done. Is the efficiency higher or lower? Does the efficiency percentage meet the minimum local standard?

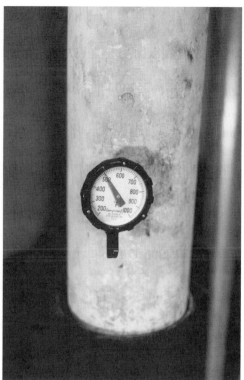

Figure 19-18 The stack thermometer inserted to take the temperature of the flue gas leaving the boiler. *Courtesy: Bacharach, Inc.*

Figure 19-19 When the stack temperature stops rising, deduct the boiler room air temperature to get the net stack temperature. *Courtesy: Bacharach, Inc.*

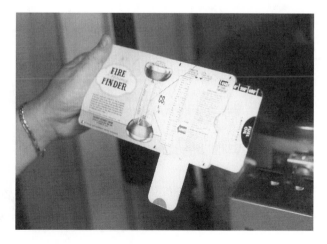

Figure 19-20 Using the Combustion Efficiency Finder to determine the unit's combustion efficiency. *Courtesy: Bacharach, Inc.*

DETERMINING COMBUSTION EFFICIENCY BY TESTING FOR OXYGEN IN THE FLUE GAS

An alternative method of determining combustion efficiency is to test for oxygen in the flue gas. The presence of oxygen indicates that there is excess air in the combustion process. The Lynn Model 7600 (see Figure 19-21) is an electronic unit that uses this method. A probe is placed in the flue pipe (see Figure 19-22) and connected to the unit. This connection allows us to take flue gas samples to determine the smoke, stack temperature, oxygen and carbon monoxide—if any is present—in the flue gas. The efficiency test is done in the same sequence as the CO_2 test. The test results as well as the percentage of combustion efficiency are all displayed on the digital screens. The oxygen testers can be calibrated by taking a sample of the air and setting the scale at 21%. Once this is done, the flue gas sample can be taken and the percentage of oxygen in the gas recorded. The theoretical magic number for the oxygen test is 0%, which would indicate the complete absence of excess air. As explained above, this result, if it could be accomplished, cannot be the final setting. Some excess air is required to ensure that the burner will operate smoke free over a long period of time.

Regardless of the testing equipment used, the test sequence should remain the same. It is absolutely necessary to prepare the unit for testing by thoroughly cleaning it first. Conventional burners require the proper –.02″ over the fire draft to allow the burner to supply the air necessary to complete the combustion process. Newer, high efficiency units that utilize forced draft burners do not have this same requirement. Once the draft is set, the smoke test must be done because any oil burner can smoke and soot-up the inside of the boiler. Soot buildup will insulate the heat exchange, raise the stack tempera-

Figure 19-21 **Lynn Model 7600 Combustion Analyzer measure the oxygen in the flue gas to determine combustion efficiency.** *Courtesy: Lynn Products Company.*

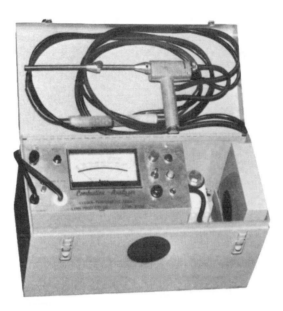

Figure 19-22 **Lynn Flue Gas Analyzer with probe.** *Courtesy: Lynn Products Company.*

ture, and lower the overall efficiency. This can happen on old conventional units as well as new high efficiency units. Regardless of the type of burner, a maximum smoke reading of #2 is the absolute limit. Anything higher will build up soot in a few weeks and all your hard work goes for naught! The CO_2 or oxygen test will reveal how well the burner is performing the combustion process. This test is also an indication of the presence of excess amounts of air moving through the boiler. Excess air removes heat that should be absorbed by the boiler and sends it up the chimney. The stack temperature is the report card in the test procedure. The lower it gets, the higher the efficiency will be. Table 19-1 shows how excess air affects combustion efficiency and the relationship between oxygen and carbon dioxide in the flue gas.

The use of test equipment cannot be overemphasized. Test kits should be part of every service technician's equipment. The cost of fuel oil and the competition of other fuels has changed the role of the service technician dramatically. Setting oil burners by "eye" just doesn't cut it any more. The scientific approach is the only way to ensure the maximum utilization of the fuel and the resulting satisfaction of the homeowners.

TESTING FOR CARBON MONOXIDE AND OXIDES OF NITROGEN

In an attempt to improve air quality, local building codes may require specific test results for carbon monoxide, (CO) and oxides of nitrogen (NO_x) in the flue gas generated by oil burners. These requirements, as well as the installation of CO detectors in many homes,

FLAME RETENTION BURNER

EXCESS AIR	% O_2	% CO_2	NET STACK TEMPERATURE—DEGREES FAHRENHEIT										
			320	340	360	380	400	420	440	460	480	500	520
15.6	3.0	13.4	86.7	86.2	85.8	85.3	84.8	84.3	83.9	83.4	82.9	82.4	82.0
18.7	3.5	13.0	86.5	86.0	85.6	85.1	84.6	84.1	83.6	83.1	82.7	82.2	81.7
22.0	4.0	12.6	86.3	85.8	85.3	84.9	84.4	83.9	83.4	82.9	82.4	81.9	81.4
25.5	4.5	12.3	86.1	85.6	85.1	84.6	84.1	83.6	83.1	82.6	82.1	81.6	81.0
29.2	5.0	11.9	85.9	85.4	84.9	84.4	83.9	83.3	82.9	82.3	81.8	81.2	80.7
33.2	5.5	11.5	85.7	85.2	84.6	84.1	83.6	83.0	82.5	82.0	81.4	80.9	80.3
37.4	6.0	11.1	85.5	84.9	84.4	83.8	83.3	82.7	82.2	81.6	81.1	80.5	79.9
41.9	6.5	10.8	85.2	84.6	84.1	83.5	83.0	82.4	81.8	81.3	80.7	80.1	79.5
46.8	7.0	10.4	84.9	84.4	83.8	83.2	82.6	82.0	81.5	80.9	80.3	79.7	79.1

CONVENTIONAL STYLE BURNER

EXCESS AIR	% O_2	% CO_2	NET STACK TEMPERATURE—DEGREES FAHRENHEIT										
			320	340	360	380	400	420	440	460	480	500	520
46.8	7.0	10.4	84.9	84.4	83.8	83.2	82.6	82.0	81.5	80.9	80.3	79.7	79.1
52.0	7.5	10.0	84.6	84.1	83.5	82.9	82.3	81.7	81.0	80.4	79.8	79.2	78.6
57.6	8.0	9.7	84.3	83.7	83.1	82.5	81.9	81.2	80.6	80.0	79.4	78.7	78.1
63.3	8.5	9.3	84.0	83.4	82.7	82.1	81.4	80.8	80.1	79.5	78.8	78.2	77.5
70.2	9.0	8.9	83.6	83.0	82.3	81.6	81.0	80.3	79.6	79.0	78.3	77.6	76.9

Courtesy of The Boiler Institute

Table 19-1   **Combustion Efficiency Chart for #2 Fuel Oil.**

have made these tests necessary. The obvious time to do the testing is either during the annual tune-up or if there is an alarm sounded in one of your customers' homes.

The presence of CO in the flue gas is an indication of an incomplete combustion process. Not enough air is supplied to completely burn the fuel. The result is smoke and soot and CO formation. It is hard to believe—but possible—that the homeowner did not call for help to solve the problem of a smoking chimney and soot all over before the level of CO in the building sounded the alarm. Follow the procedures outlined in Chapter 16, Service Procedures, Chart X. Perform an efficiency test with an electronic efficiency tester that can indicate the level of CO in the flue gas. There may be a form that has to be completed to show that the problem has been corrected.

NO_x in the flue gas is the result of excess air combining with nitrogen. This happens when nitrogen combines with oxygen at high temperatures. Usually the homeowner will call and complain of an oil odor in the building, and most of the time this problem can be solved by servicing the fuel delivery system parts. A new nozzle of the correct size, spray pattern, and angle can do wonders for eliminating NO_x. There are testers available to determine NO_x levels. You may have to buy one if your local code requires specific acceptable levels of NO_x and you have to provide the test results.

ANNUAL FUEL UTILIZATION EFFICIENCY (AFUE) RATING

The Department of Energy has established a test procedure for new oil-fired heating equipment to determine the AFUE to give consumers a guide when purchasing new heating equipment. It is similar to the miles per gallon rating found on new cars. Remember, the results of the AFUE test performed under carefully supervised conditions in a laboratory may not be achievable in the field.

The test includes the normal combustion efficiency procedure as well as other factors such as stand-by heat loss through the jacket insulation and heat lost up the chimney between heating cycles. Some manufacturers have developed packaged residential units that include an indirect water heater. At the end of each heating cycle, the hot boiler water is circulated through the heat exchanger in the water heater to produce domestic hot water. This procedure eliminates the heat loss between cycles and increases the AFUE, but leads to another problem. By cooling the boiler down after every cycle, we create a cold start situation. Cold starts produce condensation until the boiler warms up beyond the dew point of the flue gas. The condensate formed is an acid that attacks the boiler and flue connectors unless they have protective coatings to prevent damage.

The struggle to attain the highest possible efficiency for the equipment we service is continuous. When you run an efficiency test and the CO_2 reading is 14%, you will undoubtedly get a rush of satisfaction, but do not leave the burner at that setting! If you do, over a period of time as the fan picks up dust and the air supply is reduced, the burner will begin to smoke. This will soot-up the heat exchange surfaces and lead to a messy service call. The prudent thing to do is to add excess air to the combustion process that will drop the CO_2 percentage in the flue gas to around 12%. This will allow the burner to operate for a longer period of time, hopefully to the next annual tune-up, without having to be serviced.

To Briefly Summarize

1. The service department of a fuel oil company is responsible for the maintenance of the homeowner's equipment.
2. Efficiency test kits should be part of every service technician's equipment.
3. The service technician must be able to use the testing equipment and evaluate the results of the tests.
4. The testing instruments should be used to make final burner adjustments whenever the burner is serviced.

5. The efficiency test results should be matched to the results of tests taken when the unit was brand new if they are available.

Please Answer These Questions

1. Explain the function of the draft regulator.
2. What is the effect of low stack temperature on the flue breeching and flue pipe?
3. Why does the boiler have to be cleaned before the efficiency test can be done?
4. Explain why low CO_2 and high stack temperatures seem to go together.
5. Why is it necessary to set the burner to a #2 smoke or less?
6. What does "net stack temperature" mean?
7. Explain why the sequence followed during efficiency testing is important.
8. Explain how the Fyrite Flue Gas Analyzer is prepared for a test procedure.
9. How can an oxygen tester be calibrated?
10. What is the theoretical best result of the oxygen test and exactly what does this indicate?
11. An oil burner service technician performs an annual tune-up on a unit and his test results indicate a 14% CO_2. The technician then opens the air shutter and the next test indicates a 13% CO_2 in the flue gas. Why was this done and is it a good idea?

IMPROVING COMBUSTION EFFICIENCY

The previous chapter explains the testing procedure for oil-fired heating equipment. The equipment the average service technician works on will vary from an old converted coal boiler or furnace to a new high efficiency unit. Visiting trade shows is highly recommended as a good way to keep up with changes in technology and to see some of the newer units. It will be hard for you to believe that the state-of-the-art equipment at the show was how your local soot factory looked when it was new. The manufacturers of heating equipment, boilers, furnaces, hot water heaters, and oil burners invest heavily in research and development programs to improve the operating efficiency of their equipment. Many of them run training seminars for service technicians or have field representatives who can come to job sites to deal with problems. Once the equipment is installed, hopefully according to the manufacturer's instructions, it is up to the service technicians to keep it operating properly. This responsibility is an important part of service work. It is absolutely necessary to test new equipment and record the results. This information should be left on the job for future reference. If the testing is done by experienced technicians, and the best possible results are achieved, they can be used as a benchmark every time the unit is serviced in future years.

Improving the operating efficiency of a heating unit requires a complete understanding of the combustion process. The oil burner must mix the fuel oil with a sufficient amount of air to ensure complete combustion. Oil burners are rated over a range of firing rates. The air delivery system of the burner can actually deliver approximately 20% more air than is required for the top firing rate. This excess air is often used to eliminate smoking, which can soot-up the interior of the flue gas passages and insulate the heat exchanger. Try to visualize a single droplet of oil as it leaves the nozzle. It becomes surrounded by air and is ignited by the ignition spark. The droplet burns from the outside in toward its center. There must be a continuous supply of air moving with the droplet to ensure complete combustion. The trick is to accomplish this with as little excess air as possible. Excess air will increase the volume of flue gas, raise the stack temperature, and reduce the percentage of CO_2 in the flue gas.

Improving the efficiency of an oil-fired heating unit can be accomplished by adjusting the burner to get as close as possible to the theoretical 15.3% of CO_2 in the flue gas without allowing the oil burner to smoke. It is extremely important to have the equipment as clean as possible before attempting to make the necessary adjustments. Installing a new nozzle every year will ensure that the oil is broken into the tiny droplets required for complete combustion. This is true because the orifice and the distributor slots of the new nozzle will be perfect, resulting in the proper spray of tiny droplets. Older nozzles, in which the orifices may be out of round or scratched, cannot break the oil up as well. Installing a slightly smaller nozzle, and at the same time increasing the pump pressure, will also result in better atomization of the oil. The spray pattern and angle must match the air delivery pattern of the oil burner. Remember that single drop of oil as it travels from the nozzle into the combustion chamber. If it is not aimed into the air stream leaving the burner it can actually starve for air and not burn completely. The nozzle to air delivery combinations discussed earlier in Chapter 6 are good starting points for the nozzle selection process.

THE NOZZLE APPLICATION TEST

To determine the best nozzle for a particular installation, it is necessary to do a nozzle application test. The time spent—usually not more than an hour—will pay big dividends for years to come. The best time to perform this test is when the burner is brand new. The manufacturer's instructions will probably include the type of nozzle it recommends for its burner. Start with that nozzle and run an efficiency test after the burner has reached steady state operating temperature and record the results. Suppose that the nozzle had a solid spray pattern and the CO_2% at 0 smoke was 11%. Try changing to a semi-solid spray pattern and run the efficiency test again. Was there any improvement? Try another nozzle. This time change the angle and run another test. Are the results better or worse? Try a smaller nozzle and increase the pump pressure to maintain the firing rate. Do another test. Better or worse? This procedure should be followed until the best test results are achieved. Record the exact nozzle size, spray pattern, angle of spray, and test results on the information card left with the homeowner. The information will be available to future service technicians so that the equipment can be maintained properly and deliver efficient service over a long period of time.

An oil burner's best friend is a good chimney that can provide the over the fire draft that allows the burner air delivery system to supply the air necessary for complete combustion. The information supplied by manufacturers of oil-fired heating equipment always includes the chimney dimensions required for the equipment. The industry appears to be moving toward higher efficiency units that can supply forced draft that allows the units to be vented without the conventional chimney. The transition from conventional to forced draft equipment will take a considerable amount of time because of the cost involved to individual homeowners. The process may be accelerated by a government-sponsored program such as a tax credit, but that is only a possibility and not certain to occur. Because most of the equipment in the field at this time requires a chimney, you have to be able to recognize the symptoms associated with chimney problems. Figure 20-1 illustrates some common ones.

By this time, you should have been working on some oil burners either in a shop or actually out in the field. The condition of the burner drawer assembly is a good indication of how well the chimney functions. What does it look like? Is it full of carbon and does it look somewhat burned? Are the electrodes black and either cracked or crazed? Is the retention head carbonized? All of these are indications of poor chimney operation. Check the service record to see when the burner was serviced last and what work was done. If the work was done recently and has to be done again, you can be sure that there is a chimney problem. The first step in correcting chimney problems is a thorough cleaning of the boiler flue gas passages, the flue pipe and breeching, and the chimney itself. Once everything is clean, you can start to look for the cause of the problem to see if you can correct it.

Perhaps the best place to start is with the dimensions, which include the height as well as the inside area and shape. The minimum height for the smallest burner is 15′. Another consideration in the height is that the top of the chimney must be higher than the top of the building by at least 3′. This will ensure an unrestricted flow of air across the open top of the chimney that will create a form of draft known as "current draft." The chimney of a short building, which is next to a taller building, must be extended above the taller building for the same reason. There may be natural obstacles such as tall trees or hills that can restrict air movement or actually direct air down into the chimney. It may not always be possible to extend the chimney high enough to ensure the establishment of current draft. These installations must include some form of draft inducer and damper arrangement to prevent service problems with the oil burner.

The inside dimensions and the shape of the chimney are just as important as the height. The minimum size recommended for the smallest burner is $8 \times 8''$. The best shape is round, which means that to equal the internal area a round chimney would have to be 9″ in diameter. Square corners due to either a square or rectangular shape create eddy pockets that reduce the draft. If the interior area is decreased, the height must be increased to maintain the draft. If you think that a chimney is not adequate for a particular installation, you can always ask the equipment manufacturer's field representative to check it out with you. Inadequate chimney problems can be solved with mechanical draft inducers and dampers.

The materials used in the construction of the chimney are of considerable importance as well. Older buildings may have chimneys built of brick or stone. These chimneys are easily affected by cracks in the mortar that allow air infiltration into the chimney. Cold air entering the chimney will adversely affect the "thermal draft" that is the result of the hot flue gas moving up through the chimney. This problem can be detected by a careful examination of the outside of the chimney. Signs of soot inside the building where the chimney passes through can also indicate a leaking chimney. The solution to this problem is to repair the mortar by pointing the loose spots. An alternative solution is to install an

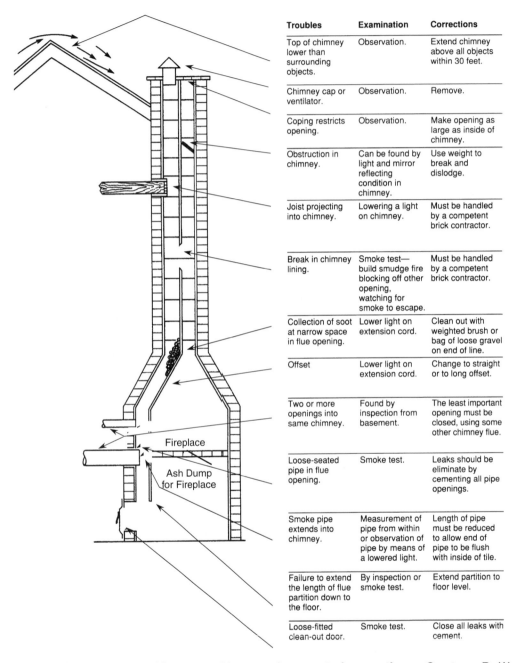

Troubles	Examination	Corrections
Top of chimney lower than surrounding objects.	Observation.	Extend chimney above all objects within 30 feet.
Chimney cap or ventilator.	Observation.	Remove.
Coping restricts opening.	Observation.	Make opening as large as inside of chimney.
Obstruction in chimney.	Can be found by light and mirror reflecting condition in chimney.	Use weight to break and dislodge.
Joist projecting into chimney.	Lowering a light on chimney.	Must be handled by a competent brick contractor.
Break in chimney lining.	Smoke test—build smudge fire blocking off other opening, watching for smoke to escape.	Must be handled by a competent brick contractor.
Collection of soot at narrow space in flue opening.	Lower light on extension cord.	Clean out with weighted brush or bag of loose gravel on end of line.
Offset	Lower light on extension cord.	Change to straight or to long offset.
Two or more openings into same chimney.	Found by inspection from basement.	The least important opening must be closed, using some other chimney flue.
Loose-seated pipe in flue opening.	Smoke test.	Leaks should be eliminate by cementing all pipe openings.
Smoke pipe extends into chimney.	Measurement of pipe from within or observation of pipe by means of a lowered light.	Length of pipe must be reduced to allow end of pipe to be flush with inside of tile.
Failure to extend the length of flue partition down to the floor.	By inspection or smoke test.	Extend partition to floor level.
Loose-fitted clean-out door.	Smoke test.	Close all leaks with cement.

Figure 20-1 **Common chimney problems and suggested corrections.** *Courtesy: R. W. Beckett Corp.*

insulated chimney liner as described in Chapter 4. This can be particularly effective when a new oil burner with smaller chimney requirements is installed.

Modern buildings have chimneys constructed of terra-cotta tile linings encased in brick, blocks, stone, or a combination of these materials. The terra-cotta tiles are either round or square and are made in two-foot lengths. These tiles are set with mortar between the lengths to seal them. The exterior brickwork is done as each length of tile is installed so that the entire chimney is sealed from the bottom to the top. You can easily spot this type of chimney because the top, which extends above the building, will be an exposed portion of one of the tiles. Remember, the top of the chimney must be wide open with no caps. An advantage to this type of construction is that the chimney remains warm and is ready to provide the proper over the fire draft without a warm-up period. This avoids problems when the burner starts on a call for heat after being off for some time.

A newer system has been developed that employs a double thickness of metal pipe that contains insulation between the layers. This type of chimney is made in two-foot sections that lock together to form the chimney. Particular attention must be paid to supporting these chimneys as they rise through or against the building. Care must be taken to use the proper shrouds and flashing that will keep the chimney away from flammable building parts. The same size requirements can be met by this type of chimney and the insulation will keep it warm between burner cycles.

Direct vent systems (through the wall) do not have these over-the-fire-draft requirements. There are two types of direct vent systems. The first type has a draft inducer (see Figures 20-2 and 20-3) either on the flue pipe or in the external wall mounted unit. The

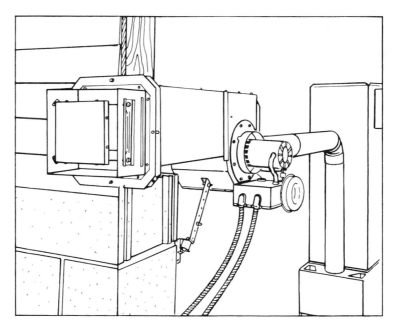

Figure 20-2 **The SideShot SS-1™ by Tjernlund removes flu gas from the boiler and brings combustion air into the building.** *Courtesy: Tjernlund Products, Inc.*

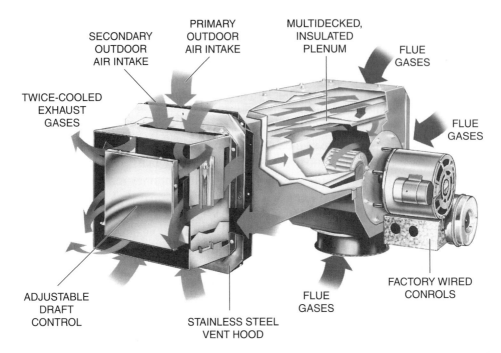

SECONDARY OUTDOOR AIR INTAKE

PRIMARY OUTDOOR AIR INTAKE

MULTIDECKED, INSULATED PLENUM

FLUE GASES

TWICE-COOLED EXHAUST GASES

FLUE GASES

ADJUSTABLE DRAFT CONTROL

STAINLESS STEEL VENT HOOD

FLUE GASES

FACTORY WIRED CONROLS

Figure 20-3   **An inside look at the Tjernlund SS1 System.** *Courtesy: Tjernlund Products, Inc.*

sequence of operation for the burner firing cycle has the draft inducer start on a call for heat. When the draft sensing switch in the flue pipe feels the draft, its switch closes and sends power to the oil burner primary control. This ensures that the necessary draft is established through the unit before the flame comes on. The second type of direct vent system has no moving parts (see Figures 20-4 and 20-5). These systems use a forced draft burner and a program type primary control with both pre- and postpurge cycles. The sequence of operation for a firing cycle has the burner motor run for 15 seconds to establish air flow through the unit. Then there is ignition and the oil valve opens to establish the flame. At the end of the cycle, the oil valve closes, the flame shuts off, and the motor continues to run for 15 seconds to clear any residual flue gases from the unit.

The heating unit chimney must be isolated from any other chimney in the building. The fireplace cannot be connected to the oil burner chimney. If there is another unit such as a hot water heater in use, it can be connected to the same chimney if it is large enough to accommodate both burners running at the same time. If this is not the case, an isolation relay and damper can be used that will keep the hot water unit off when the heating unit is in operation. Additional parts such as stack dampers may be needed on some installations but they will add to potential service problems. The best system is the simplest system.

Too much draft can lead to reduced efficiency by drawing excess air into the heating unit. Excess air increases the volume of gas passing through the boiler, which results in higher stack temperatures and reduced CO_2 percentage in the flue gas. There are two main

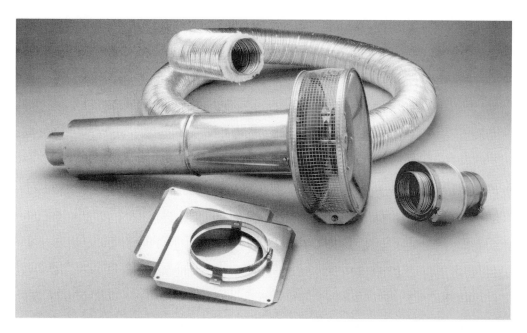

Figure 20-4   **Direct vent sidewall venting kit by Z-FLEX.** *Courtesy: Z-FLEX.*

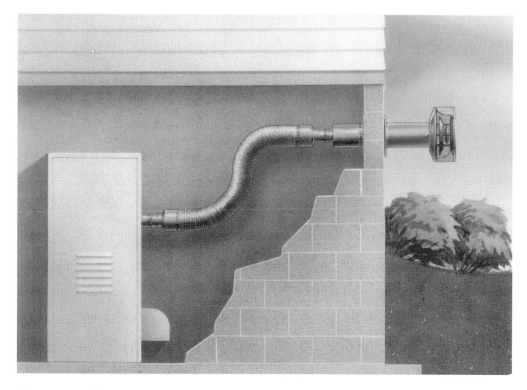

Figure 20-5   **Sidewall venting kit installation.** *Courtesy: Z-FLEX.*

areas where excess air can enter the unit. The first is through the burner where excess air may become necessary to eliminate smoke. Remember, the nozzle spray pattern has to match the air delivery pattern of the burner. When this is done, the excess air required to eliminate smoke is kept to a minimum and should not be a problem. The second and more serious problem is air entering through leaks in the unit. Openings between boiler sections, between the boiler base and the sections, around boiler doors, and around the base of the boiler are all potential areas of air infiltration. They all have to be sealed with furnace cement to ensure that no air can get into the boiler except through the oil burner air delivery system. Infiltration leaks can be detected by taking a CO_2 reading over the fire and then another in the flue breeching. A significant difference in CO_2 percentage, lower in the breeching, would indicate air infiltration into the boiler.

A properly installed draft regulator can maintain the required over the fire draft of $-.02''$ WC. The draft regulator cannot increase the draft because it allows air to enter the flue pipe, which will actually reduce the draft over the fire. The regulator can be installed either on a horizontal or vertical section of flue pipe. It should not be installed in a tee that is used instead of an elbow. A good rule would be to install the tee for the regulator close to the midpoint of the flue pipe run. The outer face of the regulator must be perfectly vertical and the damper axis must be perfectly horizontal to ensure proper operation. Use a level when installing or adjusting a regulator to be certain that the unit is installed correctly. Sheet metal screws should be used to hold the regulator securely in the flue pipe tee.

Adjusting the draft regulator requires the use of a draft gauge. Run the burner to warm up the chimney and then insert the draft gauge tubing through the boiler door so it extends over the flame. Adjust the damper weight and set it so the draft gauge reads $-.02''$. Shut the burner off and leave it off for about five minutes; then start it again. Take another draft reading to see if it is still the same. Tighten the damper weight in place so it cannot move and change the setting. Be sure that the damper gate moves freely so the draft gauge reading remains steady during the firing cycle.

Once the boiler is sealed and the draft is set, we can concentrate on ensuring that all the oil that enters the combustion process burns. The oil burners are almost 100% efficient in performing the task of burning fuel oil. It is hard to believe there are installations in which some of the oil can actually pass through the flame and not burn completely, but this does happen. Consider the fact that we are attempting to mix a liquid fuel with air that is a gas. No matter how well the fuel delivery system provides atomized fuel, each droplet is still a liquid. In order to completely mix the air and oil, we must convert every droplet of oil into a vapor. This can be accomplished by surrounding the air–oil mixture with enough heat to vaporize all the oil. A well designed, properly sized combustion chamber can provide the heated environment necessary to convert all the oil into a combustible vapor. The use of insulating soft firebrick or ceramic fiber combustion chambers is highly recommended for small oil burners. These chambers become almost instantaneously hot on the inside surface, which leads to high combustion efficiency even in short running cycles. Precast hard brick chambers are made for wet base boilers in which the insulating qualities of soft brick or ceramic fiber are not desirable. Precast hard brick chambers will heat up quickly because the wall thickness is only 1″. This allows the entire chamber to get hot in spite of this material's ability to conduct heat. A full chamber is recommended although many boiler manufacturers are installing only a rear target wall in their wet base boilers.

Every oil burner installation instruction booklet includes a chart of combustion chamber dimensions for the nozzle sizes the burner can handle. The flame must fit into the chamber without impinging or hitting the floor or walls. When this happens, some of the oil does not burn completely and a carbon deposit will build up. These deposits, known in the trade as "clinkers," can block the flame, causing smoke, soot, and eventually a burner shut down. Filling in the corners of the chamber to eliminate dead air spaces and sealing the chamber with high temperature cement to eliminate air infiltration will lead to higher CO_2 percentage readings and increased efficiency. The use of a corbell on the rear chamber wall will ensure that the end of the flame will receive the heat necessary to ensure complete combustion. A corbell is made by installing additional refractory material on the top of the back wall that extends in toward the front of the chamber. Care must be taken not to block too much of the top of the chamber with 3–4″ being the maximum width of the corbell. You can actually see the effect of corbelling a combustion chamber by the color of the flame. Most burner flames will be streaky with variations in the color and possibly traces of smoke at their tips. When a corbell, either in the rear or along the sides of the chamber, reflects heat into the flame, the result is a solid golden yellow flame with no streaks of different colors in it. When you have a flame that looks like that, you can be quite certain all the oil entering the combustion process is being vaporized and will burn completely. Over-corbelling will reflect heat onto the nozzle when the burner cycle ends. This will make the oil left in the nozzle adaptor drip through the nozzle orifice into the blast tube of the burner. After-drips of this type can create annoying odors and oil stains around the burner.

Improving the operation of the burner is only part of the process of improving overall efficiency. Attention must be paid to the oil burner's partner, the boiler or furnace. The heat released by the combustion process must be transferred to either the water in a boiler or to the air in the plenum of a warm air furnace. This exchange process can be monitored by taking the stack temperature. An abnormally high reading indicates a "sick" unit. The average warm air furnace will operate with a stack temperature of approximately 400–550°F. Reducing the nozzle size in this type of unit will control the stack temperature while increasing the length of the running cycle. Longer running cycles are desirable because they lead to better combustion performance by the oil burner. Care must be taken not to reduce the nozzle size too much because we still need enough heat energy input to supply the Btus required to heat the building. Reducing the stack temperature too much, below 350°F, can lead to condensation forming in the flue pipe, which will cause it to rot out. New, high-efficiency units are equipped with plastic flue gas outlet piping to eliminate the possibility of flue gas rotout. These connections include a drip line to drain off condensed flue gas products.

Stack temperature of boilers depends on the heat exchange that takes place in the flue gas passages. The heat released by the flame in the fire box is carried by the flue gas through the boiler and into the chimney. Modern, well-designed boilers restrict the flow of flue gas through small passages that force the hot flue gas to make contact with a large surface area of the boiler. This will extract the maximum amount of heat from the gas and result in reduced stack temperatures. Older boilers, including those originally designed to burn coal, have large spaces for the flue gas movement to the chimney. It is possible to improve the efficiency of these units by installing baffles in the flue gas passages. A few strategically placed hard firebricks can force the hot gas to make contact with the fireside surface of the boiler. This will allow additional heat to be extracted from the gas, lowering the stack temperature. Check the over the fire draft after installing baffles because they will reduce the draft, which might lead to creating smoke and the enemy of boiler efficiency, soot.

A layer of soot as little as 1/8″ thick can insulate the fireside of the boiler and raise the stack temperature. Because of this, it is essential to adjust burners with the testing equipment and never to leave a burner with a smoke reading higher than #2. Reducing the nozzle size slightly to increase the length of the running cycles will lead to more complete combustion. Longer cycles allow the combustion chamber and chimney to heat up to the steady state temperatures that create the proper environment for complete combustion. Care must be taken, particularly with steam heating systems, not to reduce the Btu input too much. Longer running cycles are desirable but the water in a steam boiler must be converted from a liquid to a gas, which eats up Btus. The indication that the nozzle size in a steam heating system is too small is the lack of steam pressure buildup during a running cycle. Another consideration in selecting the nozzle is an internal tankless coil for domestic hot water. High combustion efficiency results are wonderful when the building is receiving the heat and hot water it needs for the comfort of the people living there.

Elevated stack temperatures can also be the result of dirt buildup on the waterside of the flue gas passages. Steam boilers are particularly susceptible to this problem and an aggressive water conditioning program is recommended. This would include instructions to the owner of the building describing the proper flushing procedures of the low water cutoff and the boiler. There is no guarantee that the owner will comply with your suggested program but you really have to try. The water in your geographic area may not be a problem but there are many variations in water that can have an adverse effect on boilers. A reputable water testing laboratory can test your water and recommend the chemical treatment best suited for the area. Avoid the urge to dump a commercial cleaning compound into a steam boiler without testing the water first. You may be creating more problems than you want to have to handle. Be sure to check with the boiler manufacturer before using any commercial water treatment. Some of the newer boilers have neoprene gaskets between the sections that can be damaged by the chemicals.

The overall energy efficiency of a building contains factors other than the heating system. The insulation of the walls and roof of the building, storm doors, and windows, and the number of windows and their orientation to the sun will affect the building's efficiency. As an oil burner technician, your responsibility does not stop with the heating system because the average homeowner looks to you for advice on the overall energy picture. Many companies have expanded to include other areas of energy management.

It is beneficial that you become familiar with other related areas to widen your personal knowledge base and to increase your chances for a successful and rewarding career in this essential industry.

To Briefly Summarize

1. Oil burner technicians must be able to bring each unit up to its maximum operating efficiency.
2. You must always be aware of the combustion process and how your adjustments can affect the operation of the oil burner.
3. Experience is essential to fully understanding burner–boiler operation so that you can make changes to improve efficiency.
4. Do not stand still in the industry. Continue your education so you can become a complete heating expert and reap the rewards that are available to you.

Please Answer These Questions

1. How can reducing the nozzle size improve the efficiency of a heating unit?
2. When does reducing the nozzle size too much become a problem?
3. How does a soot buildup inside a boiler affect the operating efficiency?
4. Explain why a properly designed combustion chamber can increase the operating efficiency.
5. How does air infiltration into a chimney affect the operation of the oil burner?
6. Explain the two types of natural draft.
7. What does the term "corbell" mean and how is it used?
8. How does dirt buildup inside of the water jacket, or "waterside," affect the operating efficiency?
9. Explain the procedure for flushing a steam boiler.

INDEX